Analog Medicine - A Science of Healing

Adopting the Logic of Quantum Mechanics as a Medical Strategy

By

Ronald L. Hamm DVM

ISBN: 1-4107-1041-6 (e-book)
ISBN: 1-4107-1042-4 (Paperback)

Library of Congress Control Number: 2003090215

This book is printed on acid free paper.

Printed in the United States of America
Bloomington, IN

1stBooks – rev. 04/07/03

ACKNOWLEDGMENTS

My mentor, Doctor Grady Young is the role model I emulate and hope to become. He was the most caring, sharing, and unselfish individual I have ever had the pleasure to meet and work with.

My editors Rebecca Clack, Carrie Hudson, Virginia Galizia, Teal Bosworth, Dean Skablund, and my wife Ann were all indispensable and must share some of the blame.

I must also thank my intuitive friends Val Huber and Robin Viehweg who provided many valuable and critical insights along the way.

CONTENTS

ILLUSTRATIONS and GRAPHS

NOTE

A student of mine complained that I contradict every thing I say. This is true and you may find the same thing in the following chapters, because I am attempting to explain an analog reality with a binomial language. The Chinese Zen philosophers are famous for asking enigmatic or paradoxical questions they do not expect answers for (koans). They are simply demonstrating the limitations of language. Language is only an abstract of reality and therefore can not represent it exactly. I have made frequent use of analogies to demonstrate points. Analogies are almost like the thing I am describing, but not quite. In the end, the best analogy must always be qualified. "The Tao that can be told is not the true Tao. The name that can be named is not the constant name." *VII 13

You will also find the same thing described in slightly different ways in different chapters, because no one description completely captures that whole reality. Each description is from a slightly different perspective, and at least three different perspectives are required to describe accurately a three dimensional board.

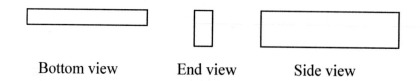

Bottom view   End view   Side view

The first six chapters are a logical progression and need to be assimilated in order. The logical conclusion of those first chapters is elaborated upon in the rest of the book in no particular order.

A fundamental problem facing the educated today is an overwhelming explosion of data. Every field is overrun with so many technical details that an individual can not possibly keep up with all the new findings and developments. The result has been segregation. Experts in one field of study seldom keep up on the latest advancements in other fields. They have become proficient in their own isolated specialty at the cost of missing out on the big picture. Even the most competent medical specialists are not usually qualified to practice general medicine. So it is no surprise that they do not have time to keep up on the latest developments in theoretical physics and logic. Unfortunately healing is a comprehensive, multi-dimensional, process that can not be fully comprehended from a strictly physical perspective. Restricting our focus to the technical details of physically suturing up a wound or splinting a fractured bone is not the way to understand healing. The secret to healing is only revealed to us when we step back and take in the more inclusive perspective. It is part of that "big picture" modern science and medicine have lost sight of.

My discussions of cloud formation, horse training, meditation techniques, martial arts, and the Big Bang may not seem, at first, to have anything remotely to do with each other or the healing arts. The basic laws of nature that determine the healing process are comprehensive and evident throughout all of reality. I suggest that you read all the chapter summaries first, to get an idea about how it all eventually fits together to produce that big picture.

After wading through all the complicated details in my proof and the accompanying analogies, we come to a final conclusion: "Healing is logical and simple."

SUMMARY OF THE SUMMARIES

1. We experience reality as eleven or more nested dimensions.

2. Healing is the lower dimensional remanifestation of a higher dimensional implicated plan or organization pattern.

3. Understanding is recognizing a familiar logical pattern within a given data base.

4. A logical format only relates the subsets of one particular dimension, and Quantum Mechanics is a logic that applies specifically to the higher implicated dimensions.

5. Healing is therefore best described, understood, and manipulated within the format of Quantum Mechanics.

INTRODUCTION

For the last thirty-five years, I have been a student of the healing arts. This book is a record of what I have discovered so far. My strictly scientific studies eventually led me to what most would call a "spiritual conclusion." However, it is the obvious logical conclusion that follows from what is now accepted as the scientific facts concerning the nature of reality. I now believe those facts show that healing is actually a re-manifestation in physical explicated reality of a higher dimensional implicated plan for the individual.

The Iowa State veterinary college, where I studied and worked as a medical illustrator, emphasized the scientific approach in medicine. That approach was limited, however, to the logic and science of Newtonian physics. My instructors were in fact strictly mono-logical by choice and design. By the time I graduated, I knew some details about the physical phenomena of healing, but the how and why were never addressed. I would eventually realize that the process of healing was actually technically impossible to explain within the Newtonian paradigm. Later on, my experience in practice proved that the Newtonian scientific approach was effective and practical as far as it went, but out in the real world it did have some well-defined limitations that often left me without recourse. In far too many cases with a patient in the field, I found myself with no definitive scientific diagnosis and therefore no logical treatment options. To begin with, the Newtonian approach did not suggest any way to directly influence the actual healing process other than to turn it off with steroids.

Another problem that bothered me was the way these scientists discredited or ignored any data or circumstances that they could not explain in the terms of their limited Newtonian worldview. The statement, "There is no scientific proof," invariably meant that they simply refused to look at the facts. For those of us who did look, however, acupuncture, homeopathy, meditation, prayer and many other alternative protocols produced obvious and undeniable results in the hands of experienced practitioners. And even more importantly for me, they directly influenced the healing process.

Ideally, I believed that scientists should be inquisitive, open-minded individuals looking at all the facts to find the truth. My professors were acting more like regimented, dogmatic technicians defending a particular point of view and logic. Their science had become a religion.

My subsequent studies of the different alternatives gave me the tools I needed to successfully manipulate the healing process, and I was never again left without something to do for a patient. The accomplished alternative practitioners that I met all seemed to be gifted with special sensitivities or innate abilities. Consequently, they could not explain logically how their methods worked or instruct an insensitive like myself on how to develop their particular talents.

My constitutional type, and years of professional education, had addicted me to the feeling of understanding that logic and the scientific approach provides. Understanding is not simply memorizing the individual steps of a particular protocol. It is recognizing a formal logical format in the raw data collected from reality. Therefore, if I did not understand the alternatives, I obviously was simply not using an appropriate logic form.

Fortunately, the physical sciences long ago developed other logical formats when Newtonian physics proved insufficient for their needs. Relativity, Quantum Mechanics, and Super string theories are all now well-established formal logics. These theories, and the realities they describe, have been the subject of several excellent books written for the 'nonmathematical' general public in recent years.

Because the description of reality in these theories goes beyond the limited range of our physical senses, any serious investigation of reality at those levels must take into account the basic functions of the brain and how it processes data. We can never separate the reality out there from our mental constructs of it; thus we have to decide finally if the falling tree in the forest makes a sound when there is no one there to hear it. The observed is defined ultimately by the observer. Knowledge of the neurophysiologic processes involved is ultimately understanding how we understand. Again, this has been the subject of many fine books written for laymen.

My studies of theoretical physics and neurophysiology convinced me that Quantum Mechanics was the best logical system to describe and explain the healing process. It provides the how and why I need to say that I finally "understand" both healing and what is going on in the different alternative modalities. Understanding is the first step in prediction, and that ultimately provides us with a greater measure of control. My subsequent successes in dealing with difficult medical conditions like cancer and allergies support my thesis. It works.

My basic "constitutional type," according to Chinese medicine, is

Earth. Earth people are said to be pensive. We put a great deal of emphasis on rational thought and are often driven to analyze things to death. In fact, energy imbalances in Earth types are often caused by excessive analysis and reflection. * I-3

Logic should not be confused with actual reality however; it is only an abstract way of representing specific parts of it. Consequently, perfectly logical conclusions are not necessarily true. Logic by definition relates the different parts of a reality to each other. We can logically consider and consequently understand only one specific dimension of reality at a time. The unity of a total reality is by definition beyond logic and therefore our understanding. Christ emphasized, again and again, that our salvation comes from belief, not logical understanding.

The logical, pensive Earth type is balanced in Chinese medicine by the intuitive, spontaneous Wood personality. Those with a Wood element constitution are more comfortable dealing with a total connected reality than the Earth personalities are. They are frequently skeptical and suspicious of abstract reasoning and prefer the unexplained, unexamined and mystical aspects of reality. The magic of ritual appeals to them, and they prefer to manipulate reality in The

Way of the Wizard. * I-3, IV-2

In Oriental philosophy, the extreme polarities of any particular quality are balanced against each other to achieve a true or healthy state.   Neither the Earth nor Wood is the preferred or correct approach; they are both elemental parts of the total and ultimate truth that can never be known.

In the why, how, and what of reality, the Wood types typically focus on the "what" while the Earth types focus preferentially on the "how".   If you are frustrated by the Wizard's complicated and ritualized "what" approach to healing and more inclined to take a logical scientific path to understanding the "how," as I am, read on.

A CAUTION

True healing may not be what you or your client/patient actually wants or is seeking.   The bulk of what we do in conventional western medical practice does not address the actual healing process. Patients/clients in our culture are accustomed to looking for and expecting only a physical fix, a cure or symptom suppression.   They are assuming a passive role, expecting the doctor or medicine *per se* to do something to or for them so that they can go back to life as usual. *I-1

Healing, on the other hand, is something that patients are directed to do for themselves. They must actively participate in the process, and that is not always easy and or convenient. In a typical healing crisis, the symptoms are actually intensified, and a major detoxification often produces new and unpleasant symptoms.

Healing is a holistic phenomenon that manifests as basic changes in the physical, emotional, mental, and life style activities. A healed patient is a different reorganized individual who has learned to deal with the diseased condition by changing on many levels. A healed patient, in other words, has learned to "be" healthy. The healed Mumps patient will be physically immune by way of antibodies to a future challenge. Another individual learns to avoid activities that result in injuries like carpel tunnel syndrome. The type "A" personality can learn to meditate to circumvent the ravages of high blood pressure and cardiovascular disease. Unfortunately, patients/clients often refuse to, or are unable to, change those emotions, thoughts and activities that contribute to or are the cause of their particular disease condition. Rolling Thunder, a famous Native American medicine man in Nevada, will put off treating patients who come to him with a schedule. Because he knows that the hectic life style of westerners is often at the root of their problem, he will only pursue a "healing" when they stop looking at their watch.* I-2

A healing doctor simply teaches patients to select an easier, healthier way of manifesting their basic spiritual or implicated plan. The process of healing originates in and is fueled by the higher implicated dimensions of the patient's reality. That reality is best described by and effectively manipulated with the principles and logic of Quantum Mechanics. Therefore, I maintain that Quantum Mechanics, is the logic of healing. That is essentially what this book is all about.

Ronald Lee Hamm DVM   April 2002

SUMMARY

In veterinary college I eventually came to understand that the Newtonian scientific format used exclusively by conventional western medicine could not logically either explain or deal with the actual process of healing. To become a healer I would have to also study logic, theoretical physics, neurophysiology, and a number of alternative medical protocols.

SCIENCE or ART

Once upon a long time ago, in a Stone Age culture far away, two tribes came together for a hunt. In camp that night, celebrating their success, a young man named Sci witnessed a truly magical event that would ultimately change his life and mankind forever. One of the members of the other tribe was a fire maker. He was a gnarly old man with no teeth. He worn a strange hat and his body was covered in mystical tattoos. The rest of his clan beat on a drum and chanted while the old man rubbed two sticks together. After a short time, smoke began to rise from one of the sticks. He laid the smoking stick in a bed of pine needles and gently blew fire into it. All in all, this was a very impressive display and also a very complex ritual.

Later, back at home camp, Sci and his friends tried to copy to the best of their collective recollection what the fire maker had done. They tried several times to no avail. They had not yet developed logic, so the only tactics available to them were memory, imitation and trial and error. They were faced with the formidable task of reproducing the complete scenario that they had observed. Every failure could be attributed to an almost infinite number of possible

1

oversights. Maybe they did not get the tattoos or the chant right. What direction was he facing when he blew fire into the pine needles? He probably has a fire God. Did anyone notice what kind of wood he used? Some suggested that they steal the magic drum. After numerous failures, they lost interest and decided it was hopeless.

Their confusion made many in the tribe feel vulnerable and threatened by this magic. The fire maker was obviously a powerful demon or witch and they suggested that he should be eliminated for the safety of the tribe.

Then, one day, word came to them of a different fire maker in another tribe close by. They decided to go and see if they could figure out what they were doing wrong. But to their dismay, this fire maker was completely different. He did not even know about the other fire maker and he said that he learned to make fire from a vision. If Sci and his friends wanted to watch, they would have to pay him. He was a young man with teeth, no tattoos and he wore a feathered headdress. To begin with, he prayed to a fire God and made an offering to the four directions. Next he struck two rocks together while his helper played a flute. He then carefully and slowly blew fire into some dry moss.

They had witnessed two different fire makers using two entirely different rituals. Now, our young man was completely confused and overwhelmed. Mastery of a complex task such as fire making without logic becomes an individualized accomplishment, and some will just have a knack for it. The fire maker is an artist, and he is personally valued by his society for his unique abilities. After all, no fire maker means no fire.

The act of fire building will seldom be advanced by the artistic approach because there is simply no incentive for the master to simplify and economize the ritual that he identifies with. He has a vested interest in maintaining the status quo by keeping it complex and magical. After all, if everyone in society could make fire, he would lose his leverage and status. However, he will give you some magic rocks and teach you the ritual-for a price.

Then one chilly morning, Sci awoke, stretched, and rubbed his hands together to warm them up. He took notice that the faster he rubbed, the warmer they get. He also knew that heat is associated with fire. A pattern began to take shape in his mind, and he had an idea. Logic was born. Maybe rubbing sticks together really fast makes fire!

With this new way of thinking about fire making, he could reduce the complexity of the situation to one factor: friction. He could now concentrate all his attention and efforts on manipulating the sticks. His subsequent success in producing fire proved that the logical approach was, in fact, an effective way to solve problems.

Sitting by the fire one night, our new fire maker discussed his great achievement. He said that he now "understood" fire making, and he communicated the idea to the others in his group. Within a very short time, almost everyone in his tribe learned the technology of fire building. Not only had they learned a valuable skill, they had acquired a new and far more effective way of dealing with life and solving problems.

Logic simplifies reality by focusing our attention on only one of the many possible series of simultaneous events or patterns going on: 1,2,3; A,B,C; Red, Orange, Yellow, Green, Blue; Friction → Heat → Fire.

Logic is an abstract idea, and the transfer of ideas between individuals is called communication. It is the idea, pattern or logic that is stored in the brain as memory, and when we perceive this same series of events in reality again, an association is made with that idea, and we say that we "understand." The logic gives a sense of meaning to the events that we observe. Understanding means that we are able to accurately predict and prepare for the future, and this ultimately leads to a feeling of control and a sense of security. We all like that.

Rituals are the reproduction of an activity without any logical basis or understanding. Rituals are enduring because there is no basis for changing their sequence. They are performed and copied, but not communicated because they have no abstract idea or meaning to be transferred. They are complicated by the lack of focus and, therefore only a few individuals in a society will have the time or ability to master them completely. Logic transforms ritual into technology. Technology is simple and economical by comparison, so many more people will be able to exploit it successfully.

Technology is the backbone and definition of culture and the basis of its evolution. We define, evaluate, and name cultures for their technological developments: "Hunter-gathers," "the Stone-age," "the

Bronze-age," "the Industrial-age," and "the Information-age" are labels we have used to describe different cultures. An advanced civilization has well developed technologies that economize on man's efforts in regard to his physical survival. Spiritual focus and development occurs only when the immediate concerns for physical survival can be temporarily set aside.

This all seems to be pretty straightforward until we realize that there is an inherit limit to the logical process. Our young Stone Age hero enjoyed a great deal of prestige as a wise sage and respected teacher. He had, after all, developed the technology of fire building, and society would always be grateful for this gift, but he soon found that people expected him to solve other problems for them with this new way of thinking. Buoyed up by his past success, he told everyone that they could also make fire by rubbing two rocks together. Just imagine his embarrassment. He had now inadvertently discovered that his logic only applied in specific cases. To explain and understand the second fire builder's technique, Sci would have to come up with another idea or logic.

The logical investigation and manipulation of reality that Sci developed eventually became known as "science." Motivated by the sting of embarrassment and a grant from the newly formed science foundation, our Stone-age hero analyzed the second fire builder's ritual. Sci conducted several experiments over the years and came eventually to the conclusion that some types of rock contained fire. He found that by striking them sharply with another rock, he could release small parts of it. Logic proved to be successful again and that salvaged Sci's reputation.

The tribe now had two entirely different and effective logics for making fire, and over time, they became polarized on this issue. The rock fire builders and the stick fire builders both believed that their method was superior and obviously dictated by God. They eventually turned to Sci and his logic for an answer. A second grant was forthcoming, and it funded another series of experiments. In the end, Sci managed to offend both groups by concluding that the two systems were equally valid. Both logics had advantages and limitations. When it rained and the wood got wet, the rock method worked best, but they could not always count on finding good quality fire rocks. Even more importantly, he found that he could not effectively combine or mix the elements of the two different logic forms because they were exclusive systems. They could use either one or the other, but not both. Striking two wooden sticks together did not produce fire. Rubbing two rocks together did not produce fire. Striking or rubbing a rock with a stick did not produce fire. He concluded that the intelligent thing to do was to use the system that seemed to be the most appropriate for the conditions at hand. Debating which way was the best was unproductive and a waste of time. Sci subsequently received the first ever "Noble" prize for this discovery.

Doctors in the twenty-first century find themselves in a similar situation. Unfortunately, our story happened once upon a long time ago, before the written word was invented. Our hero's experience was never recorded so that we could learn from it today.

Western medicine has understood the importance of the logical scientific approach. Their biomedical researchers have exploited Newtonian logic with a great deal of success in the past. They have in fact elevated the basic principles of that logic to a religion and are now content to rest upon their laurels. Why question success? Stick with a winner. There is now overwhelming evidence that their strict adherence to the eighteenth century logic of Newton has exhausted its potential. Today inappropriate, complicated, and paradoxical solutions are evident throughout the system of modern western medicine. This would suggest, as Sci pointed out, that other logic forms should now be explored. Isolated in its own specialized world, the conventional medical establishment is unaware that the rest of the scientific community has moved on to make use of the formal logics of Relativity, Quantum Mechanics, and Super-string theory. Too bad they never heard about our Stone-age hero and his Noble prize winning discovery.

On the other hand, the doctor of today is confronted with a whole spectrum of illogical alternative medical modalities. The founders of these other medical applications realized in many cases the limitations of a strictly applied Newtonian logic. But instead of replacing it with another logical format, they simply abandoned the logical process altogether. These modalities are for the most part practiced today as magic rituals, and as our Stone-aged hero found, rituals do not integrate well. They are complicated and difficult to learn because they represent the full spectrum of a specific aspect or part of reality. These modalities also isolate themselves because they cannot be logically communicated to either the scientific group or the other alternative groups. Without logic, there is no way to accurately communicate and integrate the results and discoveries of any one modality with another.

You don't have to study alternative medicine long to realize that terms like "energy," "remedy," "allergy," and "disease" do not translate between the different alternative modalities or the sciences. Each modality and science has a different definition for these terms. They represent or refer to different subsets of reality. Communication and comprehension quickly collapse when there is no common vocabulary.

Inevitably, a new "profit" appears on the scene and changes one of the fundamental steps in the ritual, and a whole new modality with a different name and vocabulary is born. This, incidentally, always happens just when you thought you were getting a handle on the old one. The masses initially flock to embrace this new improved protocol only to find eventually that it does not live up to the enthusiastic "profit's" hype, at least not in their hands. In time, one of the old fundamentalists eventually resurfaces to preach for a return to his favorite classical application. What's a doctor to do?

EXAMPLE: Let us examine the reality of a bag of one hundred marbles. We can define the two different measurable qualities of size and color. Each quality suggests a logical way to classify marbles into subgroups so that they can be analyzed. So, for instance, we could say there are ten big marbles in the bag. If I carefully define "big" in terms of a specific measure, the subset "Big marbles" will be consistent and always contain the same ten marbles. If I decide, on the other hand, to divide them by color, I can again have a definitive group of twenty "Green marbles". Both are two perfectly good ways to logically divide up the marbles. But like the two fire building techniques, the two systems cannot be integrated. Big marbles and Green marbles are not exclusive categorizations. In some cases the two subsets overlap. For example, a big green marble would be in both groups or subsets. The Big subset cannot be compared or related to the Green subset logically because they actually include some of the same marbles. You cannot relate something to itself logically.

So, if I try to relate the subsets "Big marbles" and "Green marbles" as separate elements in a logical statement, it will automatically conclude in nonsense. In this case, if we add the ten big marbles to the twenty green marbles, we will not necessarily get thirty marbles. While the two systems are incompatible, their separate logical conclusions are not. The statements that there are ten "Big" and twenty "Green" marbles are uncontestable facts that in no way conflict.

EXAMPLE: The word "Chi," from classic Chinese acupuncture theory, usually translates into English as "energy," and that often creates a great deal of confusion. Newtonian philosophy does not recognize the important subtle energy components that are an integral part of the more inclusive Oriental concept. Westerners are constrained therefore to think of the "Chi" phenomena as a more limited electrical or ionic process. This mistakenly leads them to conclude that the meridians, along which the Chi is said to move, must be physically materialized structures like nerves or vessels.

The alternatives that I have studied have not restricted themselves to any kind of logical format. There is no way of predicting which logic will be used in each successive step of their protocol. They are poly-logical, jumping indiscriminately back and forth between the principles of Newtonian physics in one step, and Quantum mechanics or Relativity in the next. Rote memorization and ritualization are the keys to mastering them.

THE BAG OF MARBLES PARADOX

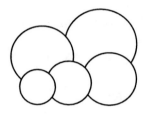

THE SIZE DIMENSION
20 Tiny
20 Small
20 Medium
20 Big
20 Large

Logic relating marbles in the size dimension 20T + 20B+ 20L = 60

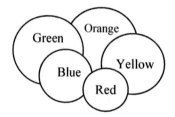

THE COLOR DIMENSION
20 Red
20 Blue
20 Green
20 Orange
20 Yellow

Logic relating marbles in the color dimension 20R + 20B + 20G + 20Y = 80

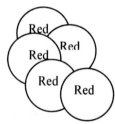

MIXED DIMENSIONS
Includes twenty five subsets
5 Medium red
2 Small green
3 Big green
Etc.

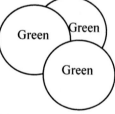

Logic relating marbles in combined groups 5MR + 2SG + 3BG = 10

Paradox may occur any time you attempt to mix subsets from different dimensions in a logical statement. For example:
   Logically 20 + 20 + 5 = 45   but   20 Green + 20 Large + 5 Medium red = 40
That is because 5 of the green marbles are also large marbles. These five marbles are part of two different subsets being added or related, and you can not logically add or relate something to itself.

EXAMPLE: Muscle testing or kineseology is a good example. A substance's <u>energy</u> profile is typically analyzed by comparing its affect on over-all muscle strength. Substances that test energetically positive are then usually taken orally in their <u>physical</u> or chemical form. A substance's energetic profile and its chemical profile are subsets of two different dimensions of reality and cannot be logically related (Big and Green marbles.) Homeopathy has demonstrated that many effective energy remedies are chemically poisonous. Kineseologists simply hedge their bet by avoiding the testing of known poisons.

The application of the logical scientific approach to the many different complicated alternative modalities will simplify and reduce them to technologies based on logic and understanding. Quantum mechanics is the logic that I believe is needed for the transformation of those rituals into scientific technologies. Unlike rituals, technologies are subject to change and evolution. Only the fittest technologies eventually survive because the <u>best</u> way to do something inevitably becomes the <u>only</u> way. The application of a strictly logical scientific approach to the presently complex and confusing situation in the healing arts will ultimately reduce it to a few basic effective technologies.

The rest of the scientific community has already demonstrated that the conclusions of different logical systems can be easily integrated to solve complex multi-dimensional problems. For example, it requires the logic of Relativity to predict that and how an atomic bomb can be made, but the logic of Newtonian physics is required to physically build it. In the same way, the materialistic Newtonian technologies of Western medicine can be effectively combined with the Quantum energy technologies of the alternatives to produce a more holistic medical format for the future. That format will ultimately be a more logical, simple, humane and a less expensive approach than the system in use today.

SUMMARY

Logic simplifies and expedites problem solving. It transforms complicated rituals into technologies. Because different technologies are logical, their results can be effectively integrated and communicated by way of well-established transformation adjustments. Technologies change over time as new and different logical formats are developed. A culture is preserved in its rituals and evolves by perfecting its technologies. It is time for mono-logical Western medicine to evolve into the twenty-first century by adopting the proven logical formats of Relativity and Quantum Mechanics as medical strategies. It is also time for "alternative medicine" to embrace the logical process and become truly scientific and integratable. After all, holistic medicine, by definition, must be an integrated, poly-logical medicine.

DIMENSIONS OF REALITY

Research in the field of subatomic physics has established that reality is basically and fundamentally a sea of energy. The electron and other so-called subatomic particles are not really material particles at all. They are simply manifestations of interference between different wave patterns in the field of energy. Their physical appearance is expressed in terms of probabilities.

Physicists have also demonstrated that this energetic reality is affected by the conscious attention or focus of an observer. Particles materialize within the energy field only when the energy of conscious attention creates interference patterns. Without a conscious input, reality manifests only as energy. *III-2, 3

Our five basic senses are material based and have limited ranges of effectiveness. We can see, feel, hear, taste, and smell only a small fraction of the full spectrum of reality that we now know exists.

A comprehensive description of reality, therefore, cannot be based on references to these limited sensory functions. Instead, scientists use the concept of dimensions to describe what is out there. Actually, dimensions describe how the energy of the conscious mind interfaces with the rest of energetic reality.

Theoretical physicists and philosophers believe that they can logically identify a minimum of eleven nested, or inclusive, dimensions. This means that the higher numbered dimensions include those below them. Three-dimensional reality includes the first and second dimensions for example. *III-3

We conceptualize energy in the first dimension as a straight line. In the second dimension, this energy transforms into a typical sine wave. This same reality is manifest as a coil in the third dimension. The fourth dimension adds relative motion to the picture and the three dimensional coil transforms into a spiral.

These observations closely parallel the worldview of clairvoyants, shamans, and mystics. Barbara Brennan indicates that she can focus on seven different nested layers in the human aura. It is interesting to note that these seven, plus the four dimensions manifesting as the physical body, also equal eleven total layers or dimensions. *III-1

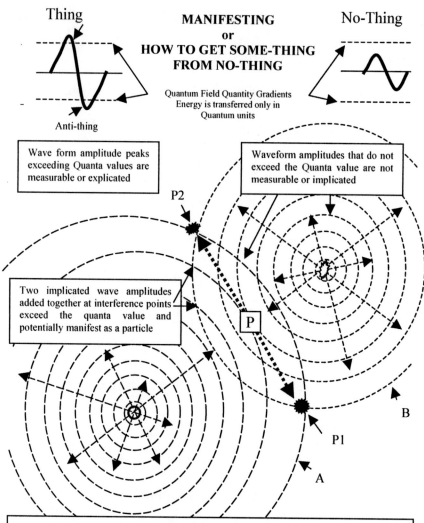

Thing

MANIFESTING
or
HOW TO GET SOME-THING
FROM NO-THING

No-Thing

Quantum Field Quantity Gradients
Energy is transferred only in
Quantum units

Anti-thing

Wave form amplitude peaks exceeding Quanta values are measurable or explicated

Waveform amplitudes that do not exceed the Quanta value are not measurable or implicated

P2

Two implicated wave amplitudes added together at interference points exceed the quanta value and potentially manifest as a particle

P

B

P1

A

A single point P will manifest when the two located transmission waves A and B first intersect. That point splits in two as the waves pass through each other. The two particles P1 and P2 fly apart on a course perpendicular to the interfering waves' transmission line.

Yin = Particle = P1 ← P → P2 = Anti-particle = Yang

The two particles technically are not <u>forced</u> apart. Force does not manifest at this dimension. They are actually sequentially re-manifesting at different locations within the interference pattern created when the two waveforms move through each other.

The mind's layering of reality is not limited to the scientists' minimum of eleven mathematical dimensions. A minimum of three dimensions, at right angles to each other, is required to establish volume and mass. But we can slice this reality into an infinite number of different planes or dimensions. Each dimension or plane manifests a different perspective of that reality.

Existence is a concept that is defined from the perspective of our five basic senses. As a general rule, most people reserve the definition of existence for those singularities that are confirmed by two or more sensory inputs. A sighting not confirmed by touch, is called a mirage or illusion, and is not considered to exist.

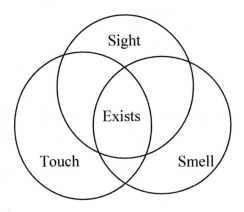

In the three-dimensional reality, sensory inputs manifest as different sized and shaped spheres of influence. The area of overlapping spheres of influence has different shapes depending on the plane of reference or dimension slicing through it. The nature or look of existence, in the same way, depends upon the plane or dimension the observer chooses to focus on. The cut surface of an orange or grapefruit only look like pie wedges if you cut it perpendicular to its axis. It appears different if you choose to focus on another plane or dimension.

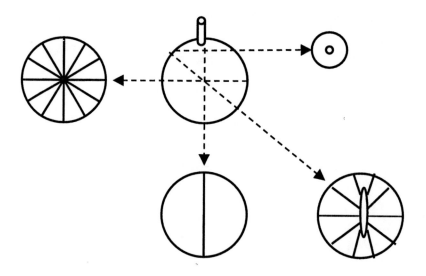

Of the eleven dimensions that scientists recognize, the physical world that we experience manifests in only the first four. There is actually more potential energy in so-called "empty space" than there is in the solid matter occupying it. The boundary between the fourth and fifth dimensions is the speed of light. Space, time, mass, and gravity of the fourth dimension are all transformed or enfolded into different manifestations of energy in the fifth plus dimensions. Energy moving at or below the speed of light, in the lower dimensions, is called **explicate**, or measurable, and manifests as matter or electromagnetic waves. Explicate energy has been described and characterized extensively by physical scientists using the theories of Newtonian physics and relativity. Energy manifesting above the speed of light is called **implicated** or enfolded energy, and is perceived as empty space. Quantum Mechanics, Super string Theory and spiritual laws best describe implicated energy. Implicated energy manifests as potential sub-atomic particles, laws, patterns of nature, thoughts, ideas, the soul, or God. *III-3

Transformation is the process of energy crossing dimensional boundaries. In the implicated dimensions, some of the energy is called magnetic. This same energy transforms into electricity when it manifests in a lower explicated dimension. The symbol for transformer in electronics is a third dimensional coil interfaced with a one-dimensional line.

Because energy in the fourth dimension moves in a spiral, we will often find spirals in nature where dimensional transformations are taking place. The spiral-shaped inner ear, or cochlea, is an excellent example. It transforms mechanical energy patterns into electronic signals.

The DNA molecule is another vivid example of a transformer. Here we have an implicated idea, plan, or "soul," if you will, traveling as pure implicated energy in a spiral or coil configuration in the higher dimensions. That energy transforms into the coiled explicated physical chemistry or matter called DNA in the fourth dimension .

Clairvoyants report spirals of energy, or chakras, entering the physical body. It is fairly obvious that these are also transformers where higher dimensional implicated energy codes are spiraling down into the lower explicated physical forms. *III-1

Transformations often result in mirror imaging or templating of the information from one level to the next. The material scientist William Tiller has physically demonstrated this "mirroring" effect. The psychic Barbara Brennan sees this same phenomenon in the upper layers of the human aura. Homeopathic remedies are based on this same flip-flop principle. They eliminate symptoms by adding to the system the subtle energy of the substances that normally produce those same symptoms physically. This would suggest that Homeopathic remedies are interfacing with the energy field at these higher dimensions. It also explains why their effects can be so powerful and inclusive. *III-1, 4

The patient must therefore be technically described as a multi-dimensional organism. The master plan for the patient exists in the higher implicated dimensions as an energetic pattern that is projected down sequentially through all the lower dimensions. The informational energy is transformed in each dimension into a different manifestation. Pure implicated energy may transform into a force in the next lower dimension and materialize in the next. The patient manifests in different implicated dimensions as spirit, mind and emotions. These non-local manifestations ultimately project down into a localized physical reality. The physical body, that is the focus of modern western medicine, is therefore, only a small part of the whole organism. The materialized anatomy and physiology that we experience in the third and fourth dimensions is only the shadow like projected image of a much larger energetic reality.

FOCUS

As the quantum physicists demonstrated, the act of focusing attention seemed to affect the reality they were studying. The energy of the mind actually determined which of the many possible forms became manifest to them. *III-2

At the level of physical reality, cause and effect is a useful concept. When we perceive a consistent series of events, the first event is said to cause the rest. If we then identify this same series in another situation, we say that we understand it. Understanding leads to predictability and the feeling of control and safety.

Unfortunately, the total information flooding our senses at any one time is overwhelming and does not reveal the simple binomial relationships needed to define cause and effect. To make some sense of this confusing situation, the brain reduces the overall input by focusing on only one dimension at a time, while ignoring the rest. It is easier to perceive relationships like 'cause and effect' and to gain insights within this limited format. Because dimensions are inclusive, the higher the dimension, the more information it contains. To solve a complicated third or fourth dimensional problem, I often reduce it to a second dimensional one, working it out on paper first.

Learning to focus means we learn to ignore. A baby learns to see a face by not seeing everything else. You don't see the forest if you focus on the tree. You don't see the deer on the road if you look at the bug on the windshield.

This same format was followed in my formal education in veterinary college. We focused selectively on courses in anatomy, biochemistry, histology, physiology, etc. We purposely confined our focus to that specific subject or dimension. The idea was that the insights we gained in each isolated area of study or dimension would be helpful eventually in dealing with a patient as a whole multi-dimensional organism. The multi-dimensional organism that we identify as our patient can have any number of individual dimensions.

Subatomic Particles	Tissues	Genus	Religion
Atoms	Organs	Race or Breed	Residence
Molecules	Systems	Sex	Nationality
Substances	Organism	Family	Occupation
Cells	Species	Tribes	Soul

I can describe a particular individual on any one of these levels. The different points of view or focus create different realities, all of which may be accurate. Inconsistencies in eyewitness accounts are due to people focusing on different dimensions in the same situation. One individual sees a male chauvinist commit a crime. Another might see an Italian do it. A third will focus on the fact he is Catholic, a Californian, or a construction worker. Each sees a different reality, perceives a different cause, and therefore assumes a different solution in the same situation. The conscious decision to focus determines what part of reality manifests for them. Jesus, for example, focused primarily on an individual's soul. A person's experience, position, or situation in life usually determines what dimension is habitually focused on in a given situation and what part of the total picture they will perceive.

An ecologist "sees" a dynamic sub alpine fir and spruce ecosystem

A small child "sees" a deep, dark, scary forest.

A naturalist "sees" an important watershed and oxygen generator.

A lumberjack "sees" a potential job and employment.

A carpenter "sees" potential building supplies.

A backpacker "sees" a camping spot and quiet retreat.

A squirrel "sees" food and refuge.

An asthmatic "sees" an allergy attack.

A preacher "sees" an example of God's work.

The story of the three blind men describing an elephant comes to mind. The subatomic physicist's preconceived idea determines what particle will manifest for him out of all the possibilities there are.

The results of focusing our attention on different dimensions of reality can be graphically demonstrated with EEG recordings of brain waves. When we focus on the lower explicated dimensions, **Beta** waves dominate. Physical activity, emotions, rational thought and language are associated with those high frequency, low amplitude waves. Specific localized electrical activity is evident. As the focus shifts to higher dimensions, the frequency decreases, and the amplitude increases progressively producing in turn **Alpha, Theta,** and **Delta** waves. In the Delta state, the attention is primarily on the higher implicated dimensions. Language and physical activity are suspended, and the electrical activity is more generally distributed in the whole brain. Maintenance, healing and being functions predominate in this state. Non-dreaming sleep, deep meditation and prayer are Delta wave states of being. *XII-4

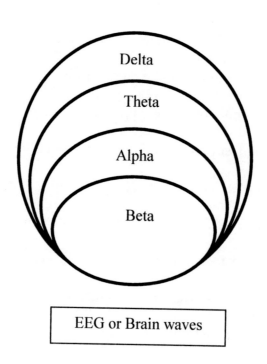

Delta

Theta

Alpha

Beta

EEG or Brain waves

From a functional standpoint, lower dimensional implicated reality is associated with the mental activities of the physical brain. Emotions and rational thought are programmed (computer-like) functions of the brain. They can be explained and analyzed in terms of the physical neural anatomy and chemistry. Consciousness, on the other hand, is a higher dimensional implicated activity that we associate with the mind. It cannot be physically explained or located.

In medicine, the dimension focused on determines what is perceived as the cause of a particular disease or condition, and therefore the ultimate course of action. In dairy herd health, infertility is a common problem.

The perceived cause of infertility depends upon the dimension of focus. It can be nutritional, infectious, genetic, managerial, traumatic or environmental, among other things physical. So it might be treated logically with vitamins, antibiotics, vaccinations, management changes, surgery, ration adjustments, genetic selection or a combination of several of the above. In this situation, the focus is on explicated, physical reality, and the elimination or neutralization of the perceived cause. I could focus, on the other hand, on a higher dimension where cause and effect is not evident. On that level of reality, an imbalance of yin and yang, a deficiency of an energetic element, or a blocked flow of Chi might be recognized. So I might treat infertility with some type of energy adjustment such as Acupuncture, Homeopathy, or Qi Gong. I have many different logical ways to treat infertility. What I eventually decide to do will be based primarily upon the dimension of reality that I focus on and the "cause" that is evident there. In the real world, medical practitioners seldom find those clear-cut, one-on-one, cause and effect cases outside of the ER. The more experience we have as a practitioner, the more dimensions we become aware of, and, therefore, the more causes we will see. The best bet is usually to include several different dimensional treatments. We can effectively combine antibiotics, vaccinations, ration adjustments, hormone injections and acupuncture to improve the fertility of a dairy herd. In fact, it would be naïve and foolish to depend upon any one solution in this multi-dimensional world.

In 1993, people were dying of an unidentified illness on the Navajo Indian Reservation. Initially, the western medical establishment was baffled as to the cause and therefore didn't know how to prevent or treat it. An Indian Medicine man named Andy Natonabah told scientists from the Communicable Disease Center that too much rain caused it. All the moisture had resulted in an unusually heavy piñon nut crop, and consequently there was an enormous increase in the deer mouse population. Years before, the Navajo had treated the same disease with ceremonies that included sand paintings of the mouse. Subsequent "scientific" investigations eventually proved that the mice were, in fact, carrying the Hantavirus, and this was causing the disease. So depending upon your dimension of focus, you could logically conclude that too much rain, piñon nuts, deer mice or Hantavirus caused the disease. *III-9

HEALING

The medical professions are supposedly concerned with achieving and maintaining health. Health is the direct consequence of healing. Healing results in health. What exactly is healing? Objective science, which deals exclusively with the explicated or lower dimensions of reality, has described the details of the healing process in great detail, but it is unable to explain the mechanism behind it. It just happens when the cause of disease is eliminated. The reason there is no explanation for healing in objective materialist terms is because it originates in the higher implicated dimensions of reality. The atomic bomb, gravity, magnetism, chi, love, consciousness, and the soul all cannot be explained objectively for the same reason.

DEFINITION: Healing is the process of a higher dimensional implicated blueprint, template, or code, re-manifesting itself in the lower physical explicated dimensions of reality.

ANALOGY: If we place a magnet beneath a sheet of paper with iron filings on it, the filings will assume a pattern characteristic of the magnetic field.

When we physically disrupt the pattern, the filings will immediately reorganize themselves (heal) the second the disrupting influence is removed. Any attempt to explain this phenomenon without reference to the implicated magnetic field would ultimately prove to be fruitless.

The healing phenomenon provides the most compelling evidence that a magnetic-like higher dimensional pattern (the aura, or soul, if you will,) provides a blueprint, template, or coded message around which the structure of the physical body materializes or manifests. The materialist, in contrast, would have us believe that some quality of the physical material in the body produces the aura and healing. They must also believe then, that some quality of the iron filings causes the pattern on the paper.

CONCLUSION: If we are to be concerned with healing, we need to focus our attention on the higher implicated dimensions of reality. The higher our focus, the more inclusive and effective it will be.

SUMMARY

A comprehensive description of reality reveals it to be a connected quantum field of energy. Consciousness is an integral part of that field. Consciousness attention layers the rest of reality into an inclusive series of nested dimensions. Physical explicated reality makes up only the lower third of these, and the higher dimensions remain implicated or un-measurable pure energy. The organizational rules or laws of nature and the master plan of an organism are part of the all inclusive, implicated upper dimensions. These implicated energy patterns project down into the lower dimensions and manifest there as physical reality. The upper implicated dimensions therefore ultimately define, fuel, and control the healing process.

IT'S ONLY LOGICAL

For an organism to survive in the physical world it must adopt a *modus operandi.* Random behavior, trial and error, conditioned response, instinct, intuition and logic are some of the possibilities.

Random behavior is a tactic used with great success by plants. It is inefficient and expensive however. Vast numbers of seeds are frequently produced for every one that actually germinates. Animals use this tactic with sperm cells.

Trial and error is somewhat more efficient. It requires a recording system like a brain to keep track of and record the trials, so that mistakes are not repeated.

Instinct and conditioned responses can be very effective if the environment and circumstances don't change. If they do, disaster is often the result.

Intuition can be effective for some, but consistent and dependable results are hard to maintain for most individuals.

Logic is a tactic that we have used to reduce the cost and increase our efficiency by accurately predicting the future. It makes use of our ability to symbolize and abstract. First, we construct an abstract model of reality and how it works (Philosophy.) Then we simply experiment with that model. The model allows us to mentally explore different options without wasting valuable resources and permits us to make mistakes without paying the price. Extended logic also allows us to explore reality beyond the limits of our sensory abilities. We can predict what is on the other side of the hill, even if we cannot see it directly. Prediction also economizes on effort by narrowing down the almost infinite number of possibilities to a select few.

Over the years, man has developed many different logic forms or systems. Each logic form begins with a basic philosophic statement or assumption about the nature of the reality being represented. This is called the "premise". The premise is actually a statement defining which dimension of reality we are focusing on. The total picture of reality is divided up into different subsets on each dimension. Each dimension presents us with a different picture puzzle. Every logic system formally defines how the individual elements or pieces of that particular reality or dimension are related. IF + AND + AND = THEN. We cannot solve a picture puzzle by mixing the pieces of two or more different puzzles. We cannot logically relate subsets from two different dimensions.

EXAMPLE: Water manifests in four different dimensional forms as ice, liquid, vapor, and gas. The logic and tools that are successfully used when working with ice are entirely different than those that would be applied to the other three forms or manifestations.

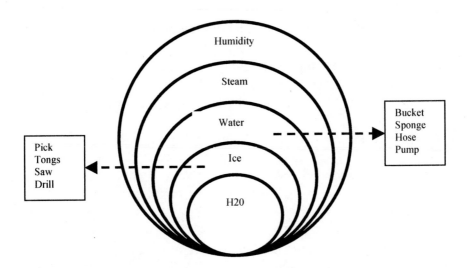

A logic form is valued and preserved if its predictions consistently prove to match those in reality. A logic form is more valued if it is simple. Complex systems ultimately require more time and effort than the trial and error method.

Mankind generally values logic positively. It is usually considered a virtue to be a rational "logical" person, especially if that individual is using the same logic we are. We feel safe in the company of rational people because they seem to be predictable to us. In the Western world, logical is usually equated with intelligence.

SCIENCE

Science is defined as the logical exploration of reality. For every logic and dimension there is a science. We have the sciences of chemistry, geology, biology, and botany, for example. Scientists are experts at using a particular logic and philosophy. They are, consequently, awarded academic degrees called "doctors of philosophy."

To be logical, we must first state a premise, select an appropriate form, and then follow each step of that particular form, in order, until we reach a conclusion. That conclusion is then compared to reality. If the conclusion contradicts the observed facts, that logic is replaced with another form. So, **being perfectly logical is not the same as being right**. Being logical means invariably following the accepted sequential steps (reasoning) dictated by the particular system selected, premise to conclusion.

Arithmetic and algebra are formal modes of logic that are used in the quantification of one-dimensional reality. Trigonometry is a formal logic form used in the second dimension. Geometry is a well-established logic that applies to reality in the third dimension.

Newtonian physics is a comprehensive logic form well established in the physical sciences. It begins with the premise that matter and energy are separate and immutable. This means that it applies to the lower four dimensions of reality where matter is manifest.

Relativity proved itself to be an indispensable tool when extreme speeds in the fourth dimension began to qualify Newton's premise. Einstein's logic began with the premise that energy and matter could be interchanged and it proved to be spectacularly successful. The atomic bomb, nuclear power and space travel were all made possible using this logic.

Quantum mechanics is a logic system that was developed by subatomic physicists. It begins with the premise that reality is totally energetic, and it applies to dimensions five and above: to implicated reality.

Super-string theory is a logic that begins with the premise that reality is entirely energetic and consists of frequencies produced by vibrating strings and the interference patterns that result. It applies to the higher implicated dimensions.

The scientists, mathematicians, and philosophers, who established the formal logics referred to above, tend to be rather isolated in their specialties from the rest of the world. Many believe that Quantum Mechanic principles are limited to the very small sub-atomic realm and Relativity principles to the very large astronomical realm. They only see their applications in those circumstances where that logic was first developed. The fact is, however, that a proven logical format is a basic statement of relationship or "law of nature." The laws of nature originate in the implicated higher dimensions as templates or plans. These project down into all the lower dimensions of reality and ultimately determine how reality manifests on each level. The informational energy of the implicated dimensions can transform into an energetic force on a lower implicated dimension and a physical thing in explicated reality. These laws permeate the whole of reality so we can recognize them in many different aspects of life.

EXAMPLE: In four-dimensional reality, Newton's third law of motion says that for every action there is an equal and opposite reaction. In the theory of electro-magnetism, energy is portrayed as having two opposite and equal poles. At the implicated subatomic level, every potential particle has an equal and opposite antiparticle. In their picture of reality, the Chinese recognize the Yin/Yang quality in everything. In religious doctrines, a Satan always balances and therefore defines God. These are all expressions of the same basic law manifesting in different dimensions.

In the last chapter, I pointed out that the dimensions are inclusive. The logics that apply to those dimensions are also. Quantum Mechanics is inclusive; it can accurately predict all the tenants of Newtonian physics and Relativity when the proper transformation calculations are applied. It is by consensus the most accurate and productive formal logic ever developed by man. It can explain and predict events in fields as diverse as psychology, ecology and the martial arts. I will show in the following chapters how it can be used to design and execute a new medical paradigm.

Man has developed many informal logic systems as well. In fact, each dimension that we can focus on will, or should, have a specific logic that applies to it. A role in life is actually a statement of premise such as; "if I am a father," "if I am a farmer," and "if I am a Methodist." My particular culture then defines the rules that apply in each of those dimensions.

EXAMPLE: Mike Tyson manifests in different dimensions as atoms, chemicals, a human, a black man, a husband, a fighter and an American, among others. The logic he applies so successfully in his professional dimension is inappropriate, obviously counterproductive, and even unlawful when applied to the other dimensions of his life.

EXAMPLE: Games are forms of applied logic. The premise is a statement that we intend to play a particular game. Let us say checkers. The logic is spelled out formally in the rulebook. The game is played with the assumption that checker rules will be followed at each and every step of the game. Gaming theory dictates that both participants follow the same rules or logic. If my opponent, on the other hand, decides instead to apply chess rules, we cannot play at all. First of all, we will not agree on how a particular piece is moved, taken, or what constitutes winning. The chess player uses both the red and black squares while, I see and use only the black ones. From my checker point of view or dimension, a red bishop does not exist.

The brain preferentially decides to perceive, by way of focus, only those aspects of reality that historically make sense to it. The logic that we habitually use determines what we will normally perceive. For example, kittens raised experimentally in a strictly horizontal world cannot see vertical structures such as a chair leg. *IV-6

USING LOGIC

Faced with a particular problem, intelligence dictates that we first select an appropriate logic form, one with a premise that corresponds to the dimension of reality we are dealing with. The details of the situation are then plugged into the logic format 'if ... and ... and ... then.' The specific rules of that form are then methodically applied, step-by-step, until a conclusion is reached. That conclusion is then compared to reality or our stated goal. If the logical conclusion seems to match, we are encouraged to follow through physically. If however, our logical conclusion and reality do not match we discard that particular logic and select another.

PERTURBATION

If no acceptable logical answer is found, we have two basic options. First, we can abandon the logical approach altogether, falling back on trial and error or intuition. Or we can adjust the inappropriate answer with a second application of logic. This is called a perturbation adjustment.

EXAMPLE: Binomial linear logic forms have been used to calculate trajectories in space travel. These forms are unable to account for more than three variables at a time. Unfortunately, space contains many bodies (gravities) or variables. So a trajectory is calculated using the three most dominate or massive bodies. That trajectory is defined as a single variable and a second trajection is calculated using it and the original fourth body. This process is then repeated in turn until all the significant variables in space are accounted for. This can obviously become a very complicated and complex calculation. (Thank heaven for binomial computers.) *III-3

PARADOX

When two perfectly logical answers contradict each other, we say that we have a <u>paradox</u>. When this occurs, it indicates that the logic being used is flawed or that it is being applied in the wrong dimension of reality.

EXAMPLE: Experiments using perfectly good Newtonian logic proved that light was both a particle and a wave. Einstein realized that this paradox indicated that Newton's logic, which was so effective in describing reality in the lower dimensions, was being inappropriately applied. This led him to question the appropriateness of Newton's premise in the fourth dimension of reality, and to suggest a new one, $E=MC^2$. *III-3

EXAMPLE: Spirituality is focusing on connectedness, unity, and non-judgmental acceptance or love. It describes the analog reality of the highest dimensions of reality. When we try to apply a binomial logic to spiritual matters, we first must divide mankind into different subsets. We have the saved and condemned, the believers and the atheists, and different religions. Binomial logic relates subsets, and that inevitably leads to comparisons, judgments, inquisitions, witch burning, and one war right after another. These are paradoxical to truly spiritual answers.

CONCLUSION: Logic forms applied to the wrong dimension of reality get complicated with perturbation and result in paradoxical and inappropriate answers.

PERTURBATIVE ADJUSTMENTS

Each Dimension Divides Reality into Different Parts or Subsets

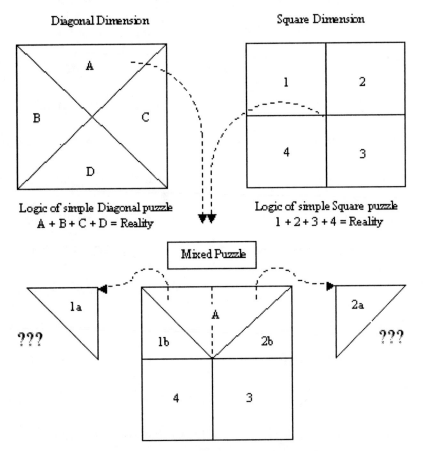

Diagonal Dimension

Square Dimension

Logic of simple Diagonal puzzle
A + B + C + D = Reality

Logic of simple Square puzzle
1 + 2 + 3 + 4 = Reality

Mixed Puzzle

???   ???

The logic of mixed or multidimensional puzzles require perturbative adjustments
Such as: Cut parts 1 and 2 in half diagonally
1 = 1a + 1b   and   2 = 2a + 2b

Adjusted Logic becomes   A + 3 + 4 + 1b + 2b − 2a - 1a = Reality

The logical solution is now more complex and the conclusion has some extra
parts or data, that do not fit in. These extra parts are just discarded or ignored.
Perturbance typically results in complexity and some enigmatic answers.

THE FISHING TRIP

We like to ride and hike the public lands of the intermountain west. To accomplish this, we must use several different maps. Maps are abstract representations of reality that we can use to figure out how to get from here to there. They are graphic logical formats.

My friend from back east will be joining us for a fishing trip into the Yellowstone backcountry. To get there he will first have to refer to an airline route map which indicates that he cannot fly directly into Grace, Idaho, from his home in Chicago. He will have to first fly into Denver and, from there, to Pocatello, Idaho, because that is where the airports are located.

We will be pulling a horse-trailer from Grace to the Turpin Meadow trailhead, just south of the park. For this part of the trip, we will need a road atlas. This map tells us that we must take route 34 north to Alpine Junction, Wyoming. There will turn east on highway 26-89, which passes through the Snake River canyon to Jackson Hole. At Moran Junction, just north of Jackson, we turn east again, traveling several miles, until we get to the Turpin Meadow turn-off. My friend's perfectly good airline map will be of no use to us on this part of our trip.

Once we get to the trailhead at Turpin Meadows, we will need a Forest Service map of the trails in the Teton Wilderness. This map shows how we can get to different places in the wilderness on hiking and horseback trails. On this horseback trip we will leave both the airline map and road atlas in the truck at the trailhead because they are of no use to us on the trail.

After we make camp on the Buffalo Fork of the Snake River, my friend wants to climb up to and fish a small un-named glacial lake above timberline. There are no established or recognized trails going to this lake, so he will have to just wing it or use a U.S. Geologic topographical map of the region. The airline map, road atlas, and forest service trail maps that so effectively got us to our base camp will be of no use to him on this climb.

We used four different maps on this successful trip to capture a Californian golden trout. Each was an accurate and useful abstract representation of the same terrain. However they each showed a different dimension or scale of it.

The logical premise in each case was the statement indicating our intension to travel by air, truck and trailer, horse or on foot. That premise determined the dimension of our focus or interest, and that in turn established which of the four maps or logics applied.

The fact that my road atlas is of no use on the trail or on the climb up to the lake in no way discredits its potential value as a map (logic) in other situations. I will be using it again, in fact, on the trip home to Grace, when my focus once more is on the highway system.

The four maps in this case were inclusive. They all described the exact same terrain. The airline map shows only the airports. The road atlas has both the airports and the highway system. The forest service map includes airports, roads and trails. The topo has all that and the topology. I could actually calculate our whole trip with a full set of the topographic maps. But if we are driving or flying, they contain far more details than I need or want. Their small scale also means that we would need a whole stack of them to cover our trip, and that would complicate the process beyond all practicality.

The four logics of Newtonian Physics, Relativity, Quantum Mechanics and Super string are logics not unlike the maps we used on that fishing trip. They all tell us how to get from here to there in reality. They all describe the same thing, but from different scales or dimensions. They are also inclusive.

The intelligent thing to do is to first determine what scale or dimension of reality we are dealing with and then select the map or logic form that applies.

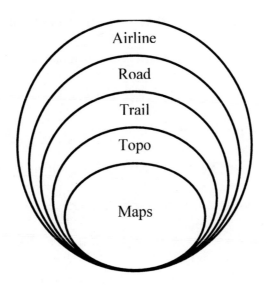

LOGICAL MEDICINE?

Doctors in this country are trained in the scientific tradition. We generally think of ourselves as being logical and scientific. In reality, we actually use trial and error reasoning most of the time. We use the techniques and drugs that have worked for us in the past. I am sold on trying something new if it is presented to me in the logic format. However, I usually do not have the technical know-how to logically deduct what a specific substance will do in or to the body of a particular patient. I simply assume that those who developed it do. After all, they are scientists aren't they? After an initial trial use of the product, I adjust the dosage and application according to my own experience with it. This is obviously an example of trial and error reasoning. I never personally used any scientific logic at all in this case.

The medical profession in this country professes to be scientific. The logic that they have chosen to use <u>exclusively</u> is Newtonian physics. They are proud to proclaim themselves to be mono-logical. They have been very successful because they realized, early on, the importance of strictly following a logical format. The double blind study, that has become the standard for their research, is designed to eliminate all elements from the higher dimensions of reality that are not covered by the Newtonian premise. Their accomplishments in this domain are truly logical and impressive.

Unfortunately, the profession has chosen to rest on its laurels. They restrict their definition of science to this one philosophy and the lower dimensions to which it applies. Rather than accepting paradox as an incentive for exploring different logical formats, they simply choose to ignore them or deny their existence. "There is no scientific evidence that......," means that there is no objective physical evidence, and they refuse to even consider any other form. "There are no red squares." The double blind studies that western medicine considers to be the ultimate proof are impossible to even construct within the formats of quantum physics or super-string theory where every thing is connected.

Physical scientists, on the other hand, long ago expanded their view of reality by embracing the sciences of Relativity, Quantum Mechanics and Super-string theory. They have acknowledged that there are both red and black squares.

After such a successful start, modern medicine now flounders under the weight of problems created by their attempts to apply an inappropriate Newtonian logic to medical problems that obviously originate outside the physical dimensions. They are attempting to put a square plug in a round hole. We see complex technology developing out of bounds as one perturbation adjustment after another is added to the mix to make the answers come out right. In areas like oncology, we find the paradox of poisoning and mutilation in a profession who begins with the premise, "First do no harm."

ALTERNATIVE MEDICINE

Advocates of alternative medicines all have one thing in common: they have perceived the limitations of a strictly material based philosophy and embrace some of the elements that follow from an energy premise. This is where the similarity ends.

When most alternative modalities were developed, no formal energetic philosophies existed, or the people who developed those modalities were not interested in such things. Without a systematic format to follow, these alternative modalities indiscriminately ramble back and fourth applying principles of Newtonian physics in one step and then some type of quantum or energetic principle in the next step.

None of the ones that I have studied constrained themselves to follow any kind of logical format at all. Because they are not logical, they are not scientific, and consequently cannot be correlated with each other or with conventional medicine. **This is not to say that they do not work**.

Each modality develops its own specialized terminology and techniques, which make absolutely no sense to anyone in the other modalities; this effectively prevents any transfer of useful information between them. So we see a growing number of different isolated modalities in energy medicine. Notice how a new one is always popping up?

Because these modalities are illogical, they become more complicated with perturbation adjustments. It usually takes a specialized expert or a person with a "gift" to get consistent results in these fields. By contrast, any competent conventional veterinarian can be expected to properly apply an antibiotic or perform a technical procedure because they are logical technologies. Given the "if-and-and" of the situation, any rational individual will arrive, basically, at the same conclusion and results.

EXAMPLE: Herbology is a modality that has been around a long time, in one form or another. It is an ill-defined modality with a wide range of philosophic approaches. On the one extreme, we have the traditional Chinese herbalists who apply remedies according to their established energetic profiles. Individuality is a characteristic of the energetic dimensions. Their original energetic focus is compromised in the second step by a strictly binomial categorization of the remedy at the genus species level. All aspens are assumed to have the same energetic profile, for example. In application the physical quantification of dose and its application are of critical importance. This is a perturbation adjustment needed to prevent materialistic, chemical overdoses that are not obvious from or have anything to do with the energy profile.

At the other end of the spectrum, we have the western herbal tradition that is basically focused on the materialistic idea of "active" chemical ingredients for treating conditions and diseases. They traditionally reject binomial, reductionism and the idea of refinement that the conventional allopathic schools preach. They favor a more analog or holistic application. Salicylic acid, for example, is said to work best in its naturally occurring compounded (analog) form. However, the focus and emphasis is still basically on the material, chemical dimension. The physical consumption of the plant material, and a binomial quantitative dosing, is an essential component of the Western Herbal system as well.

The western herbalist's basic format is compromised by the limited nature of their binomial material data bank. A perturbation adjustment is required when the active ingredients are unknown. Therefore many remedies are applied according to historic records of trial and error data with no logical format.

Realizing the inherit problems associated with this confusing situation, some herbalists have moved to energy or analog testing. They will use muscle strength testing or some other dowsing/divining technology to match up the patient with a specific herbal remedy. Unfortunately that analog energetic assessment is then usually followed by a strictly materialistic, binomial application. In other words, they consume it physically after testing it energetically.

A plant's energy profile and its chemical profile are subsets of two entirely different dimensions. They cannot be <u>logically</u> related to each other in a statement or protocol. Again I must re-emphasize; Just because these approaches are illogical does not mean they will not work for specific individuals.

EXAMPLE: Chinese acupuncture is ancient in origin and no two schools teach it the same way. It is primarily an energetic protocol and many masters logically begin with a form of dowsing/divining called pulse taking. The pulse diagnosis determines which meridians need energetic balancing. In an interesting twist, most schools switch to binomial logic to decide which points to stimulate in the treatment. They refer to several point categorizations such as transporting, command, connecting, and master points. The specific treatment points are selected from these lists, as if all patients are the same or standardized. The complete point prescription is put together according to a complicated binomial logic of proposed cause and effect energetic relationships called Traditional Chinese Medicine. T.C.M. elements are strictly implicated and cannot be confirmed or demonstrated using materialistic methods of measurement. An implicated energetic reality and the principle of cause and effect do not manifest in the same dimension and are not logically compatible.

Acupuncture treatments that are eventually prescribed are considered to be energetic adjustments. The treatment is finally evaluated physically. The question arises: Why complicate the procedure with the translation problems that inevitably rise when implicated energetic elements are substituted for explicated physical ones and vise versa. In this particular case, they must translate between every step of the protocol. The original premise in acupuncture is that we ultimately intend to manipulate the Chi, or energy, of the body. It would be logical and far simpler to just dowse/divine the specific treatment points in the first place.

Following a logical format has the very practical advantage of producing more consistent and reliable results, as well as being simple.

EXAMPLE: According to the records, Edgar Cayce, i.e. "The sleeping prophet," was consistently a most effective healer. He reportedly diagnosed and prescribed while in a self induced trance or sleep. He was obviously divining information from the implicated energetic dimensions of reality. Many of his prescriptions were rather unconventional to say the least. True to Quantum principles, his logical orientation was to focus on the <u>individuality</u> of each patient.

His cases have been meticulously recorded and cataloged by his followers in the hope that they could be used to reproduce his results. Unfortunately applying his treatments to conventionally diagnosed and <u>categorized</u> groups of patients is not logically compatible with his original approach. Needless to say, their success rate has been less than spectacular.

QUANTUM MECHANICS

People reading this book will be familiar with the tenants of Newtonian physics and relativity, so I will not waste time describing them. Q.M., on the other hand, is often referred to, but most people cannot tell you any thing about it. The mathematics involved is not relevant to our discussion here. It is really quiet simple, only a handful of principles describing an energetic reality without matter (implicated reality.)

ENERGETIC: Quantum physicists established that at the dimension of subatomic reality we have a continuous sea of energy.

Within this energy are waves, which interfere or combine with each other to produce patterns. These patterns of interference are the singularities or quantum events. The energy of a singularity is proportional to both its wave's amplitude and frequency.

ENTANGLMENT: All events are connected waveforms. There is no space or time. A measurement of any quantum affects all other quanta simultaneously. Nothing has an independent existence. The conscious observer is an integral part of the total picture of reality.

SUM OVER PATHS: All possible outcomes occur simultaneously. This idea is mathematically represented in Schrodinger's wave mechanics (equation). The act of measuring or focusing simply forces one of the possibilities to actualize. Scientists say that measurement of the system collapses the state vector, reducing the wave function to a single probability.

UNCERTAINTY: We cannot know or measure more than one property of reality. The more precisely we measure one quality the less precisely we can measure another. This is due, in part, to the fact that the measurement is usually made with electromagnetic waves. Higher frequency waves needed for accurate measures have so much energy they change the singularity. Low frequency waves cannot measure very accurately. This is called the Heisenburg uncertainty principle.

PROBABILITIES: Exact outcomes cannot be predicted. Only the probability that they will occur can be calculated. We do that utilizing Heisenberg's matrix mechanics.

NON-LOCALITY: There is no empty space and, consequently, no distance, speed, or time. All points become equal and there can be no location. There is no cause and effect because that requires time or sequencing.

This is a concept that is as difficult to explain in words, as it is to picture. In fact, we can do neither. We are trying to describe an implicated subjective reality with explicated objective language and visualizations that necessarily involve matter. However, you cannot visualize zero, infinity or magnetism either but you have probably become comfortable with these ideas. *IV 7 III 2, 3

ANALOGY: Fill the bathtub to the rim. The water inside the tub is a connected system. If we add some more water to one end, water will instantly spill out the other, but not the water we added. The effect of the added water is transferred instantly to the other end of the tub without <u>anything</u> moving across the tub. Lower the temperature below freezing and the water begins to transform into ice. Raise the temperature and the same water transforms into vapor and then a gas. The water, ice, vapor, and gas are all the same thing, H2O manifesting in different ways. In the same way energy can manifest as matter, electricity, electromagnetic radiation, work, magnetism and implicated plans in different dimensions.

Quantum mechanics describes implicated reality as an amorphous, vibrating, timeless unity. Consciousness appears to be the organizing force that determines what manifests for <u>us</u> out of that reality. In fact, Einstein said that reality is relative. Focus is the mechanism.

In implicated reality there is no empty space for a thing to occupy. Consequently, there is no distance, speed, or time. There is no cause and effect in the upper dimensions of reality.

Einstein stated that <u>no thing</u> could travel faster than the speed of light. If it does, it transforms into energy, a lot of it. In the implicated reality, events occur everywhere simultaneously. This can look like events moving faster than the speed of light, but nothing actually moves at all. It was already there. Simultaneous events cannot be framed in a cause and effect relationship.

Because everything is connected, a quantum event cannot be measured by another quantum event such as me, without being changed in the process. *III-2, 3

The most important fact established by Quantum scientists is that sub atomic particles are directly affected by focused attention and expectation. They also conclude: subatomic particles are the building blocks of all physical reality, including living bodies. It logically follows: **we can affect the physical body with focused attention and expectation as well**.

TRANSLATION: How do these principles affect the practice of medicine? They do not come into play if the focus is restricted to physical reality. Newtonian physics is most effective when we physically manipulate matter or fix. However, if the intent is to facilitate the energetic, healing response in a patient, the focus must be on the higher implicated dimensions and the following rules apply.

1. The treater cannot be separated from the treatment or the treated. Double blind studies are impossible in quantum

reality.

2. The symptom or condition cannot be considered separate from the patient as a whole. The focus is always on the individual patient (name) and his or her particular balance of energy. You cannot treat a condition or disease.

3. The patient's reality in this dimension is not confined to the physical body. The physical location of the patient in relation to the treater is irrelevant. A focused consciousness attention is ultimately the only connection of importance.

4. Doctors do not cause outcomes, they can only select or choose them. No physical effort is required, only **concentration**, **focus**, and **intent**. Instead of physically doing: imagine, let, allow, encourage, guide, or teach.

5. All influences are by additions. You positively support or alter the patient rather than subtracting or negating a condition or etiological agent. You cannot logically subtract by using negating ideas such as stop or remove. The very act of focusing on a condition like a tumor actually reinforces it because the negating part of the message is not received.

However you can effectively cancel an inappropriate frequency by adding its mirror image to it (homeopathy.)

6. The subtle energy adjustments that result from implicated energetic treatments (balancing) are instantaneous because there is only present tense. However, the explicated results that manifest as a consequence of that adjustment may take some time.

7. No quantitation is possible because every potential implicated thing is actually a mutating process. Energetic assessments cannot logically include degrees of severity. Clairsentient characterizations of an imbalanced energy profile must be framed in strictly qualitative terms. You cannot accurately sense how much pain the patient is experiencing, how long they have had it, the extent of the physical damage or predict the consequences associated with it.

8. The doctor does not determine the results of healing. Healing is a patient's individualized response and does not conform to any standard or expectation. Healing means a return to that individual patient's implicated plan or purpose. The tension

produced by resisting a programmed (normal) transition in that plan such as growing up, aging, or dying will manifest as disease.

9. Judgments are a form of categorization. There are no good results or bad diseases.

10. To directly influence a patient's or your own healing process the focus must be from the implicated mind and not the explicated brain.

The LOGIC of REMEDIES and CURES

Health care professionals the world over have turned to plants for medicine. Plants are complicated multi-dimensional organisms. In the terms of Quantum mechanics, they have several probable wave functions (dimensions) in their state vector. The premise of western medicine's binomial logic dictates that they will focus initially on the materialism or chemistry of the plant. Cultures and modalities using energetic logics, on the other hand, will focus first on its energy profile. Quantum mechanics says that conscious focus collapses the state vector, reducing the wave function to a single probability. In plain English, this means we create the reality (dimension) we choose to focus on.

The chemist, in true binomial mode, will obviously focus on the matter and then attempt to separate the plant material into parts or chemicals, because binomial logic cannot deal with more than two or three variables at a time. His obvious logical conclusion is there are only one or two active ingredients. To isolate and refine these active ingredients he has to process the plant material. This usually entails adding other chemicals and extraneous energies such as heat. Processing effectively destroys the plant's original vital energy profile. Binomial logic dictated a materialistic focus and that collapsed the state vector of the plant's wave function to that single probability. The chemist's remedy therefore eventually manifests as the materialistic one he believed it to be in the first place.

Analog cultures like the Chinese and Amerindians focus more on the plants' energy profiles. Their logic dictates that they preserve their herbs with a minimum amount of processing. They prescribe their herbal remedies according to how their vital energy will balance the patient's profile. A paradox ensues because many plants are physically poisonous. They fall back on a long history of trial and error experience and documentation (complications) to avoid these material poisonings. This is a perturbation adjustment they need because their actual remedy includes both the chemical and energetic qualities. The documented success of their plant remedies proves to them that the plant's vital energy is the critical factor in their success. They collapse the state vector of plant remedies to a different wave function and probability.

In Analog medicine, I avoid the herbalist's paradox and need for complicating perturbation by separating the energy from the chemistry. In other words, I make homeopathic plant remedies. These energetic remedies are totally analog and therefore fit logically and simply into the Quantum format. My conscious choice to focus strictly on the energy wave function of my patients and plant remedies creates the reality I call Analog medicine.

Ronald L. Hamm DVM

SUMMARY

The most efficient way to organize the sensory data coming in from reality is logic. Logic is an abstract construction of the laws of nature. It provides us with a sense of **understanding** and helps us to accurately **predict** future events. Formal logic defines the sequence of steps in reasoning or calculating. We have developed many different logic forms and they each have a very specific dimensional application. Newtonian logic is very effective in the lower four dimensions. Relativity is the logic that applies to the transition state between the fourth and fifth dimensions. Quantum Mechanics and Super-string theory are the logics that apply to the higher implicated dimensions of reality. Healing is re-manifesting a higher dimensional plan. We need the logics of Quantum Mechanics and Super-string theory therefore to explain the healing phenomena and to predict how to effectively influence it.

Quantum physics has conclusively established that subatomic particles are affected by the focused attention and expectations of an observer. Patients are ultimately made of subatomic particles.

Therefore, **doctors can <u>scientifically</u> affect physical changes in patients with focused attention and expectation**.

Probably the single biggest misconception about logic and understanding is that there is a preferable or correct one. The truth is we can never apply logic to and therefore understand the whole picture of reality. Wholeness or totality has no parts for logic to relate. The best we can do is to relate the parts of defined and isolated systems within the whole such as the dimensions of reality. The ultimate truth and understanding is "**There is no ultimate truth and understanding, only a collection of different dimensional ones.**"

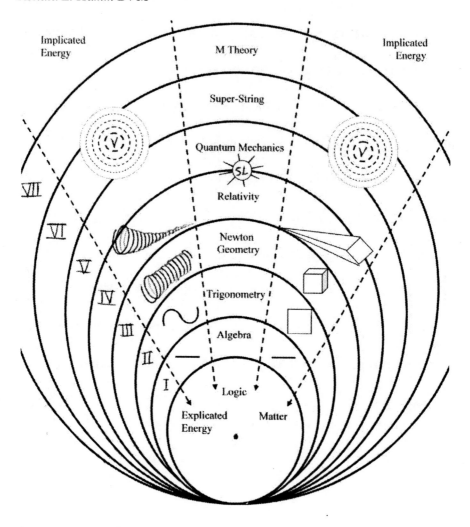

Logic is dimensionally specific		
SL = The speed of light	V = Vibrating implicated energy	Roman numeral = Dimension

THE YIN AND YANG OF NEUROPHYSIOLOGY

The concept of mental laterality is well established in western science. This basically says that there are two modes of function in the brain, the left objective side and the right subjective side. The left side includes functions of rational thought, language skills, and the focus is on material objects or the "real" world. The left uses inductive reasoning primarily. The right side 'supposedly' is more involved with feelings, intuition, and creativity. It uses deductive reasoning mostly. The most recent research qualifies this idea somewhat. The "left" mode is now more accurately described as the left-localized and the "right" mode is characterized now as more generalized or distributed. *V-8, 10

A similar dichotomy is recognized in the autonomic nervous system. The sympathetic part deals with active functions like those concerned with flight and fight. The parasympathetic part deals with maintenance activities like digestion.

In his book <u>The Body Electric</u>, Robert Becker suggests that there are two different nervous systems: An electrical one made up of neurons and a magnetic one made up of the peri-neural cells (The rest of the body). *V-1

Chinese philosophy begins with the premise of a fundamental dichotomy in reality. We call it the Yin/Yang theory. In this theory, everything is analyzed as a balance between two extremes. *V-3, 4, 5, 6, 7

In western science, we conceptualize energy as a dynamic interplay of the basic dichotomy we call charge or polarity. Electromagnetic physics is the analysis and manipulation of positive-negative charges or north-south polarities. * XV-2

I maintain that all these ideas are describing the same fundamental quality of reality, and we can put them all together into one comprehensive theory based upon the simple distinction between the analog and binomial processing modes of the brain. Simple is good.

The act of focusing is of central importance in quantum reality. As I pointed out earlier, the definition of focus involves a simple choice. We can focus on the total reality or a part of it. We can focus on one part or another. "You can't see the forest for the trees." Take your pick because the choices are exclusive. If we see the whole forest, we are in analog mode. If our attention is on a specific tree, we are in binomial mode. This simple distinction is the fundamental truth that all the theories above are referring to. Analog reality is all connected. Binomial reality is separated into parts or units. Cursive is analogue and printing is binomial.

The neuron is the basic unit of binomial mode function in the nervous system. It takes in energy information or data from the environment in analog form, builds a charge, and then fires intermittently. The basic analog input is transformed into a binomial message for transfer to the brain. The mechanisms of transmission within the central nervous system are essentially electrical and chemical. This mechanism is relatively <u>fast</u> and is used to control striated muscle functions such as the physical movements needed for doing things.

Robert Becker found that the electrical neural system was superimposed on a more ancient and pervasive magnetic system which controls those functions in the body that require slower, more sustained, long- term stimulation like healing, regeneration, and maintenance.

A comparison of these two systems reveals that the neuron based binomial system is basically electrical and is confined to physical material conduction. Information is transferred along specific nerve fibers or by chemicals called neurotransmitters. The effect or charge is perceived to move along the nerve fiber or through the blood stream.

On the other hand, the magnetic system processes information in analog mode. Magnetism projects through "empty space," which we now know is actually the implicated energy field. Magnetic fields are all-pervasive and are not constrained by specific physical structures in the body. The polarity fields are essentially static or unmoving and information is transferred by way of intensity fluctuations (vibrations) and polarity shifts.

A dichotomy begins to become evident in Becker's work.

Binomial	Analog
Neurons	Whole body
Located	Pervasive
Matter	Energy field
Fast	Slow
Electricity	Magnetism
Moving	Static

Western science has established, with EEG recordings, that the male and female process information differently. Electrical brain wave activity is usually more localized in males and it is primarily in the left hemisphere. Psychologists have also noticed some time ago that males tend to use tunnel vision, while the females are generally more multi-focal. Females apparently use more of the brain when they think. Males also focus more on the binomial words and syntax in a conversation, while females are prone to focus on the analog tone of voice and the subtle implied meanings. *V-10, VII-12 So if we add sexuality to the dichotomy list above, the male would obviously be in the left column and the female in the right one.

Anyone familiar with Chinese philosophy knows that yin is female and yang is male. They would also recognize yin qualities in the right column and yang qualities in the left. If you stayed awake through the chapters on dimensions and logic, you will undoubtedly notice that quantum mechanics describes the right column and Newtonian physics describes the left.

This dichotomy list must be viewed in the same way that the Chinese describe yin and yang. The items in the list are only the extremes of a spectrum. Few things, 'if any' are entirely yin or yang.

The majority of things in reality contain some of each quality. So, now, let us expand our original dichotomy list by adding these additional qualities and a few more to boot.

Yang	Left ← Spectrum →Right	Yin
Matter	.... light..................	Energy
Male	....	Female
Beta	.........Alpha ....Theta	Delta brain waves
Binomial	....	Analog
Temporal	....	Non-temporal
Electric	....	Magnetic
Doing	....	Being
Neurons	....	Whole body
Sensory	....	Intuition
Vector	....	Circle
Fix	....	Heal
Cause	....	Choose
Inductive	....	Deductive reasoning
Generalization	....	Individualization
Simplify	....	Complicate-details
Brain	....	Mind
Sympathetic	....	Parasympathetic nervous system
Separate	....	Connect
Words	....	Pictures
Understand	....	Believe
Active	....	Passive
Judgment	....	Acceptance
Ego	....	Unity
Newtonian Physics	....	Quantum mechanics
Categorize	....	Name
Day-light	....	Night-dark

Too big of a leap you say? Well, stick with me for a minute or two. A theory is only as good as the accuracy of its predictions. In the next chapters, I hope to demonstrate how this dichotomy list can be used to explain and de-mystify the healing process.

The mind has two exclusive modes. It can function in either the right analog mode or the left binomial mode. Each mode has its own specific mental hardware and logic. The analog mode is activated when we focus on items in the right column and the higher implicated dimensions of reality, and the binomial mode is used when the focus is on items in the left column and the lower explicated dimensions.

The fuel is a finite psychic energy, the vehicle is focused attention, and the map is our dichotomy list. Focus on the whole forest and the energy goes right. Focus on a tree the energy goes left. Focus on the whole patient and it goes right. Focus on the disease or condition, and the energy is sent left.

If I could divide my psychic energy by focusing on elements in both lists at the same time, I will have only half as much energy to use on each. The two results will also often be at odds with each other. So for example, if I eat a meal before exercising, the parasympathetic nervous system in charge of my digestion has to share the energy with the sympathetic system in charge of doing,' and I won't do a very good job of either. Remember the admonition, "Don't go swimming right after eating or you will get cramps." In the same way, if I invest energy in doing, I have less to use for healing. This is ultimately the reason for meditating. By focusing all my psychic energy in the right analog mode, which is characterized by delta wave activity in the brain, most of the energy is applied to healing and maintenance.

Healthy, successful individuals or organizations are balanced left to right. They will focus alternately, comparing one perspective with the other. Successful communication is accomplished only when both parties focus in the same mode. *VII 12

People who cannot switch are usually considered mentally ill. Autism is being stuck on the right side in analog mode. They have trouble surviving in the physical world because they simply don't focus well on the left column of our dichotomy list. The profile of an autistic personality demonstrates how our dichotomy list can be used to accurately explain and predict reality. The autistic personality profile equals the right column. *V-2

EXAMPLE:

1. Autistic individuals think in analog mode pictures instead of binomial words. Animals also process primarily in analog mode. Temple Grandin is an autistic animal scientist who is very good at designing animal handling facilities because she sees and therefore understands reality the same way animals do.

2. Dustin Hoffman's autistic character in the movie <u>Rain</u> <u>Man</u> can remember all the cards played because the right side does not judge their value and select which ones are worth recalling.

3. The autistic individual is not easily influenced by rational language and reasoning, which are, of course, binomial. You cannot easily talk them into or out of anything. In the movie, Hoffman's character repeatedly says, "I am a very good driver." When, in fact, he cannot drive at all.

4. They are overwhelmed by all the sensory information coming in because they are unable to judge the data and selectively focus on isolated relevant parts or facts.

5. Autistics are extremely dogmatic and cannot tolerate deviations from their routines. That is a defining characteristic of analog mode processing. Analog people and cultures typically do not invent. They copy, reproduce and ritualize.

A culture, society, or any organization can become unbalanced to the right as well. Fundamental religious groups, for example, create societies or governments that share many characteristics with autistic individuals. A business organization that promotes employees based upon "cronyism" rather than production is another good example.

Sociopaths, on the other hand, are stuck on the left side in binomial mode.

1. They are focused entirely from their own isolated egos and are unable to sense any kind of unity with family, country or God. Consequently, they do not perceive any obligation or need to consider anything other than to their own self-interest. They have no conscience; no sense of what society considers right and wrong.

2. They have no intuitive sense about the actual physical consequences of their actions on others. They cannot empathize with others.

3. Their narrow range of focus (tunnel vision) is always on material reality. They are not distracted by the consequences of their actions on others so they usually are very effective at manipulating the physical world to get what they want.

4. Their physical orientation and self centered perspective results in an inclination and need for constant sensory stimulation which often results in sexual promiscuity.

5. Sociopaths do not do well in small groups or with personal relationships.

An institution or organization can be unbalanced to the left as well. Conventional Western medicine is stuck on the left by choice. Large impersonal corporations tend to attract this type of employee. Governments such as the German Nazis were obviously sociopathic.

SUMMARY

The oriental concept of Yin and Yang and the western idea of mental laterality can be integrated into a single comprehensive dichotomy theory based upon modal processing. The binomial or digital mode corresponds to the Chinese Yang and western left. The analog mode corresponds to the Chinese Yin and western right. In this dichotomy the lower physical dimensions are primarily associated with male-yang- left, and the higher implicated dimensions are more female-yin-right. As I pointed out in the chapter on logic, the lower dimensions are described by Newtonian physics and the higher ones by Quantum Mechanics or Super-string theory. Therefore, the higher dimensional healing phenomenon is best understood and manipulated by applying the right-yin-female qualities in a Quantum Mechanical or Super-string format.

PROCESSING

Focusing energy into the left binomial or right analog modes is not the end of the story by any means. Once committed, the energy now has to be processed through the specific program of that mode. The programs are like two different personalities competing for a share of the finite quantities of psychic energy. They are unified in a marriage that benefits them both, but they maintain some self-interest and different points of view. *VI-1, V-8, V-10

The objective, binomial male side is basically in charge in the conscious state, but the subjective, analog female side is ultimately much more powerful. *IX-7

BINOMIAL

The objective, binomial, male program has these characteristics and functions.

    Sensory
    Controlling
    Temporal
    Analytical
    Inductive-simplifies
    Censors
    Expedites
    Organized

Judges-Moralizes
Literate
Active

SENSORY: He "perceives" or receives information directly from the sensory organs. His perspective and emphasis is largely on that reality, but he can also sense some intermittent and indirect communications from his analog side in the form of feelings and intuitive flashes.

CONTROLLING: If the left wants the energy, he has priority. We may need all the energy now to run and climb a tree because there will be no healing if the tiger eats us. If he judges that analog mode is being wasteful or irrational, he will commandeer the energy.

TEMPORAL: He perceives time flowing from the past through the present to the future. Cause and effect is his reality. His priority is to select, control, or cause the preferred future tense, and he only values the present as a stepping-stone to that future. When he thinks that present tense sensory input has enough information to accurately establish a category or label, he has it shut down or blocked. This automatic block of sensory input is the key to understanding the nature of the binomial mode and ultimately how to control disease problems caused by his unrestrained activities.

He must stop any dynamic process to analyze it with his binomial linear logic, because that logic requires stable immutable elements to relate.   This is, in fact, the premise of Newton's physics, which formally describes binomial reality. If number one changes during a count, all the rest of the following numbers in the series lose their defined relationship and meaning. For example, the force acting upon a body must have an enduring quality (immutability) if the equal and opposite reaction is to have any meaning. Binomial mode logic can only predict the state of the future tense by changing a dynamic present tense (mutable) situation into past tense where things don't change.

ANALYTICAL: He lives in the physical world where cause and effect is apparent and important. He has created a model of reality based on past experiences, and he compares all incoming data to it.

INDUCTIVE: He uses inductive reasoning to reduce overwhelming amounts of chaotic data down into generalized patterns and categories. Details → generalizations.

CENSORS: He censors the incoming sensory data so that it

matches his model or makes sense in other words. He tries to keep the brain from becoming confused by conflicting sensory data. So, contrary to common logic, looking is not seeing.

EXAMPLE: If we present to one eye a star in the middle of a graphed field, we can move it around until the star's image disappears into the blind spot of the retina or fovea. The brain inevitably perceives a whole perfect graph without any blank spot in it. We know logically if it cannot see the star, it obviously can't see the lines of the graph hidden behind it. *III-2

EXAMPLE: Kittens raised in an environment devoid of either vertical or horizontal structures will grow up not being able to perceive or relate to those qualities in reality. The kittens in the first group will run into a table leg because they don't perceive it. Those in the second group will walk off a cliff or step because they do not perceive them. They don't perceive it, if they first don't believe in it. They can't believe it, if they have never experienced it. *IV-6 "He would not have seen it, if he hadn't believed it," is therefore a more accurate version of that old, time-honored retort.

EXPEDITES: The binomial side is obsessed with efficiency. He is constantly trying to economize, so if the right is wasting either time or energy, he will intervene. Rational activities take time and energy, so whenever possible he will try to reduce routine activities to conditioned responses. The time-consuming evaluation step is replaced with a simple trigger mechanism, and the response automatically and quickly plays out without need for conscious attention and control.

ORGANIZES-PLANS: Binomial mode perceives time flowing past to future and is committed to the idea of cause and effect. Goals and purpose are the end result of this line of thinking. These give us a reason for organizing our activities in the present tense.

JUDGES-MORALIZES: He is constantly comparing data and ideas to history, his proposed model of reality, and his goals. When reality does not play out according to his model, it is bad. If a series of events are similar to one stored in memory, he assumes they will conclude in the same way. If an activity will result in less control, it is deemed immoral. His worst nightmare is loss of control. He will avoid chaos and anarchy at all costs. God must agree.

LITERATE: Language is strictly a binomial function.

Abstraction and syntax are both binomial by definition. He is inclined to talk a problem to death.

ACTIVE: He is an active, pushy, forceful, in your face type of guy. He wants everyone to believe in and therefore confirm his present picture of reality.

ANALOG

The subjective, analog, female, yin mode has these characteristics and functions:

> Extrasensory perception
> Perfect memory
> Non-temporal
> Deductive
> Dogmatic
> All knowing
> Immoral
> Literal
> Positive
> Illiterate
> Bashful
> Passive

EXTRASENSORY: Data and information is received through extrasensory mechanisms like the energy chakra. She has access to information that cannot be physically sensed. Sensory data supplies far more subtle information to her than the binomial mode is able to perceive. She also receives and responds to communication directly from binomial modes. She perceives binomial's evaluations and suggestions.

PERFECT MEMORY: She is not confined to the sensory organs limited view of reality. Realities full spectrum is actually perceived and down loaded into analog memory as pictures. If the binomial mode can be by-passed, we can access this data. All the details missed or censored out by the binomial mode are pictured there.

EXAMPLE: The 'Rain Man' character can recall all of the cards played because he saw them pictographically without abstraction or judgment. The binomial evaluates each card's worth and tells us not to remember the unimportant ones.

NON-TEMPORAL: The analog mind is present tense. It is focused on the trip rather than the destination.

DEDUCTIVE: Deductive reasoning complicates binomial's reality by emphasizing the details. It is effectively the opposite of induction. Binomial induction labels a group while analog deduction names an individual. Females tend to see and remember faces better than males.

DOGMATIC: Analog does not acknowledge cause and effect, so she cannot propose a strategy of her own. She accepts the evaluation of a binomial mode as real. This is the basis for hypnosis. Any objective or binomial mode can suggest to the subjective analog mode how reality is, and the subjective will play out that scenario, oblivious to any conflicting sensory input. She proceeds according to the original binomial input because she cannot suggest another one, on her own, to replace it. She is the perfect reproducer or copier and seldom the inventor or innovator. Cultures that are essentially analog, like the Chinese, are often characterized this way.

ALL KNOWING: Included in the perfect pictographic memory of analog reality are some basic universal laws of nature such as mathematical relationships, music and time passage. Individuals we call savants can draw extremely accurate and detailed reproductions of reality from a single observation. Others can perform extraordinary mathematical calculations, or play musical pieces without any prior objective learning experience. They are directly accessing their analog memory bank. When they subsequently objectively study these areas however, they may lose their gift because the binomial mode censors out many important and significant details.

IMMORAL: Because she is not limited to binomial's simplified version of reality, cause and effect is not as evident or compelling. Confined to the present tense, control is not an issue for her. Activities that advance a cause are moral ones. With no future goal or purpose to focus on, no casual activity is indicated, moral or immoral.

LITERAL: She takes binomial messages literally, so she has no sense of humor. Metaphors that you use repeatedly in conversations or self-talk can give the powerful analog mode the wrong message. Blasphemy and other unintentional emotional statements are literally interpreted by your own analog mode and can, in some cases, come back to haunt you. ("What a pain in the neck" for example.) Affirmations are positive messages from the binomial brain intended for the analog mind. They must be carefully worded to avoid any inadvertent miscommunications.

POSITIVE: She cannot hear negations. If he tries to send the message, "Don't buck, don't buck" to a horse, she will receive "Buck,

buck." You need to send positive statements such as, "Stand still."

ILLITERATE: She understands but cannot use language. She communicates indirectly with the binomial by way of dreams, images, symptoms, feeling, intuitions, and symbols. Information typically is sent to the conscious mind in flashes of insight without effort. In the chapter on Divining I describe how we can establish more direct communication between the two modes.

BASHFUL: The analog mode is accustomed to working in anonymity, letting binomial take all the credit and the limelight. When her messages are ignored by the conscious mind, she is inclined to stop communicating. You have to acknowledge and act on intuitive analog information if you want to continue receiving it.

PASSIVE: The binomial mode can accept analog information but cannot forcefully take it. Active attempts to pursue analog data actually block its perception. Intuitive answers typically come as passive flashes of insight after you stop trying to remember or concentrate on that particular problem.

NOTE: Analog information is stored in the quantum field as a holograph. It has qualities similar to a magnetic field pattern. It does

Analog Medicine - A Science of Healing

not move or require any conductive circuitry. Therefore no time or force (effort) is involved in analog perceptions.

Health is a balance of these two perspectives or modes. Once we appreciate how the two modes interact with each other, we can begin to effectively take conscious control of the bilateral shifts. Many disease conditions are the direct consequence of processing errors, improper modal applications, or an imbalance of application. Allergies, destructive emotional behaviors and substance abuse, for example, are all the direct consequences of an unrestrained binomial mode. His obsession with materialism, labeling, sensory suppression, and economizing often results in these inappropriate and counterproductive automatic conditioned responses. These problems are easily handled with a balancing dose of some analog perspective.

MAKING A HOLOGRAPHIC PICTURE

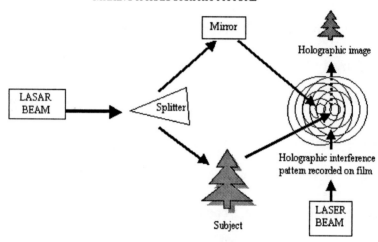

MAKING A HOLOGRAPHIC MEMORY

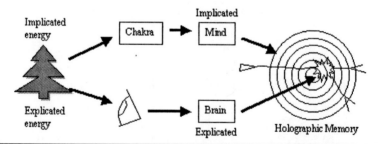

Memory of some explicated experience is stored (located) in physical circuits of neurons as reflex loops and nerve tracts. The rest is stored with the implicated memory (non-locally) in the quantum field. The unified universal consciousness is split into explicated (binomial) and implicated (analog) energy patterns and when these two perspectives of reality are interfaced, a holographic interference pattern results. Most of the data in memory is stored in that holograph. The strictly analog component of memory (intuition and insights) is non-located in the quantum field where it contacts and is potentially shared by other analog intelligences.

On the other hand, an over dependence on analog mode makes survival in physical reality almost impossible. The <u>Rain</u> <u>Man</u> character would not last long in traffic. That type of problem can be remedied by directing the patient's focus to some basic practical survival measures (binomial point of view).

Betty Edwards authored a book entitled <u>Drawing on The Right Side of the Brain.</u> It is relevant here because she demonstrates several practical ways to take control of the processing switches. *VI-2

She makes the initial observation that any realistic drawing, such as a portrait, is accomplished by working from the right side of the brain. We will concur because <u>pictures</u>, <u>individuality,</u> and <u>details</u> all appear in the right hand column of our dichotomy list.

To facilitate the shift to the right she suggests drawing the <u>spaces</u> between or around the physical structures. This shift of focus to the "empty space" worked well with objects like chairs, etc. We now know that empty space is actually implicated energy, and that is in the right column of our list as well.

When drawing portraits or other solid objects that do not include definable spaces, she suggested turning the models upside down.

This proved to be amazingly successful as well, because it avoids the left side's automatic block of the sensory input that follows labeling. We do not recognize, and therefore label, familiar structures from this perspective. When labeling does block present tense sensory perception, we fall back on the left side's memory bank or past tense. The left presents us with a composite picture of all the ears we have ever seen and labeled. That becomes reality according to the left, and so that is what our hand draws. It only generally resembles the model.

By turning the face or model over, recognition is stymied and the sensory input is maintained. The hand now draws what it actually sees in the present tense rather than the left's censored and simplified historic impression of it.

We can apply this same type of technique or ones similar to them in the practice of medicine. Acknowledging that healing is also a right-sided activity, we can deal with it more effectively by simply turning the patient upside down.

QUALIFYING NOTE:

My theory of modal processing seems to contradict some of the well established tenets of the laterality theory. In that theory, **emotions** and **creativity** are always associated with sensitive, female, artistic, and other right brain qualities. They are juxtaposed across from the left brain's rational thinking. In my theory they are binomial, left brain activities. This disparage is due to the fact that these two theories are describing the realities of two different dimensions. Laterality theory is concerned with, and effectively applies to, the complexity of sociological profiles. The modal theory, on the other hand, is describing a more fundamental and basic mechanism that is part of a higher dimensional reality.

EMOTION: Emotions are forms of the classic subconscious, automatic, conditioned reflex. Binomial judgments are the triggers for these physiologic <u>reactions</u>. They are designed to commandeer vital energy reserves to fuel a programmed physical response. Emotions are the B in an A$\rightarrow$ B $\rightarrow$ C stimulus response loop. We have:

Judgment$\rightarrow$ emotion $\rightarrow$ physical response.

This is bad → sad → grieve.

This is dangerous → fear → run.

This is morally wrong → outrage → punish.

That is sexy → arousal → X- rated.

That is denying → anger → attack.

Contrary to popular and even professional opinions, emotions are functionally left brain and binomial. As proof, we find that emotions always interfere with or completely block accurate subtle energy readings from a spouse or close friend. Also the healing effectiveness of "mindful" meditation is due, to a large extent, to eliminating the tensions created by emotions and rationalizations of the left brain.

You have also probably noticed that "love" is not listed above as an emotion. In modal theory, the sensation we call love is actually the anti-emotion. It is the "release" (see chapter on health and healing) that naturally occurs when we suspend judgments and embrace our connection. It is the sensation we feel when tension is absent. Love is passive, and does not preclude any kind of tension or physical response.

CREATIVITY: Creativity is another quality the laterality and modal theories disagree on. It is classically assigned to the right by the former and solidly on the left in the later. The creativity associated with artists is actually the ability to perceive the same reality from a different dimension or perspective. That different perspective was not created by the artist, it was always there as an integral part of holographic reality. It positively resonates with us and we appreciate it, because we intuitively recognize it as true. As proof consider the fact: individuals (autistics), organizations (religions), and societies (Chinese) that are predominately right brained are characteristically not creative. They are, in fact, dogmatic to a fault.

It is the left brain, in modal theory, which constantly explores for new and better ways to frame reality. A truly new and creative approach does not positively resonate with the older formats and we do not initially like them. They are originally perceived as unsettling and scary.

SUMMARY

To deal logically with physical/material reality we must focus in binomial mode. To deal logically with subtle energy and the healing process, we have to focus in the analog mode. Once the attention is focused in the appropriate mode, different rules of mental processing apply. We have to be cognizant of these basic differences in processing to become more effective, logical doctors.

Probably the single most important modal processing activity that we have to deal with in medicine is binomial mode's shut down of the sensory functions. This is triggered by and automatically follows the binomial acts of labeling and categorizing. Controlling this mental process is the basic principle taught in the oriental martial arts as well.

THINKING

I was just thinking about thinking. Is that actually possible? Self reflection has been a philosophic quandary since man developed language. A camera cannot actually take a photograph of itself. It can only record a reflected, and therefore, reversed image of itself. Will our attempts to understand thinking also produce a mirror image of that particular reality?

Philosophy is reality abstracted into language. As such, it has some very definite, built in, limitations. Just because a question can be logically posed and answered does not mean that it is a meaningful representation of reality. The classic Chinese philosophers were legendary for asking enigmatic questions such as: "What is the sound of one hand clapping?" They were simply trying to illuminate this problem.

I speak a form of English, and this language has cause and effect built into its syntax; the noun *verbs* the object. Cause and effect is the assumption or premise of this logic form. It describes a basic linear relationship. It is difficult or even impossible to fit elements of reality into this format that do not manifest that quality. The upper implicated dimensions are more circular than linear. If I try to compose the above question about thinking in English the noun and object are the same: Can <u>self</u> perceive <u>itself</u>? Logically one can not relate something to itself. The verb collapses into an identity or equality, *self=itself*. The perturbant mechanism we use to cover this problem in English is the verbs of being. Being is defined as a form of existence, and existence is defined as a state of being. The word *is* is simply substituted for *equals* in this circular definition. This circular logic more closely represents the form of upper dimensional implicated reality where everything is connected. Scientists avoid the complexity and confusion created by perturbant adjustments like this by resorting to a simpler mathematic format to analyze such situations.

Life is multidimensional and the phenomena that we commonly refer to as "thinking" is also. Thinking manifests differently on different dimensions.

<u>Consciousness</u> is the highest dimensional and therefore most inclusive manifestation of this process. It is often equated with "attention." It is identified with the organizing element in the chaos of

reality that many scientists refer to as the unified quantum field. God and Heaven are the terms many will use for this.

Mind is the lower manifestation of consciousness that has the ability to focus attention or choose. It is that element of thought that demonstrates free will. Intent is basically how we describe thought in this dimension.

Mental is the rational or logical level of thinking. It has to do with how the intent of the mind can be achieved in physical reality. It is the highest level of thinking that we can recognize as being reflected in the physical structure of brain tissue. Neuron A is connected to neuron B and that is in turn (logically) connected to neuron C

Instinct is a form of the thought process that is sometimes used in lieu of the mental rational process. In this case the adaptive, present tense decision process is replaced with a hardwired past tense decision.

Emotion is at that level of the thought process where the mental or instinctual plan begins to be physically executed. It is the preparation stage where the physiology is adjusted in expectation of action.

All the above dimensional manifestations of the thought process are implicated or physically immeasurable. They are like a pervasive magnetic field pattern. They are best described and manipulated with Quantum physics or Super-string theory. They do not logically fit into the linear format of the English language. Hopi might be the best language to use when discussing these.

Ions and neurotransmitters pervade the whole physical body. These begin to organize and move according to the field patterns established by the intent of the mind. Remember how the magnetic field organized the iron filings on the sheet of paper.

Neurons precognitively materialize around the potential mental and emotional thought patterns that have been used successfully in the past and consequently preserved in the genetic code.

Reflex loops are established within the chaotic mass of neurons and serve as potentials. Some loops are recognized and reinforced by use. Neurons and connections that are never used are overwhelmed eventually by entropy and disassociate into chaos.

Nerve fibers and tracts are again materialized ahead of time by genetics in anticipation of their potential need.

In this top-down scenario the implicated intention to move

something eventually projects down into physical explicated reality to become nerve, muscle, bone, and a hand.

This is also the reason why mental and emotional problems can and often do result in physical problems such as disease.

These lower dimensional physical manifestations of the thinking process can be described by Newtonian physics and successfully manipulated with that logic. They can be logically and effectively related by a binomial linear language like English.

Logically we can not relate the implicated elements of the higher dimensions and explicated elements of the lower dimension in the same statement. We can not relate chess and checker moves in the same game. We can not relate the green marbles to the big ones. If you do, the results will be paradox, complexity, and confusion. Unless you are comparing them directly to each other such as: The brain contains both the mental and emotional functions.

You can not logically deal with highly inclusive elements like consciousness or the mind within the limited parameters of Newtonian physics or the English language.

There is no space-time, beginning or end, or cause and effect on those levels. You can not discuss or even consider the location, cause or beginning of the mind or consciousness.

You can not design a double blind, controlled experiment to study them. These ideas and mechanisms simply do not compute on the implicated levels of reality. You do not order a T-bone steak at Mac Donald's, because it is just not on the menu.

Ideas are basically the mental constructions or associations we create by focusing selectively on specific aspects in the chaotic incoming data. Solidness is an idea we associate with a two dimensional plane and mass with three dimensional constructs. It is the association that is down loaded into memory as an idea. The idea in memory primes us first to believe in its existence and therefore preferentially to focus on and see this same association again. Three linearly sequential events are associated with the idea of movement. The neural pathway A –B –C, established by the first sensory experience is reinforced by subsequent experiences. The sensory data is stored in the cortex next to the motor activities that would ultimately respond to it. This forms a potential reflex or stimulus-response loop. The activation of the response part of the loop reinforces the sensory part. Therefore, the more times an idea is perceived and used to initiate a response, the more likely it will be used again. Electric currents flowing through the neural loop preserve it and stimulate its physical development.

<u>Thinking</u> is a term that many use to mean combining individual ideas or neural pathways in new and different ways. This reflecting is an internalized, active process that requires the investment of vital energy to replace the stimulus from incoming data of perception. Energy is also needed to overcome the tendency of the electric current to follow the reflex path of least resistance or the habitual path. This is the point where the theory of physical determination breaks down. The intent or decision to choose an alternate and consequently more difficult route is the manifestation of free will. Free will is an expression of the implicated mind, in the dimensions above those of the physical genetic code, material brain tissue, and the stimulus response mechanism. The mind is not constrained by the laws of nature that manifest below it in the physical realm. From this more inclusive position in the dimension above, it can alter or modify physical effects in miraculous ways. It does this not by breaking or

suspending the laws of physical nature but by adding its anti-law (or its mirror image) to it. For example we can add a centrifugal force to a centripetal force.

The process of thinking is how we solve problems. We try to logically match up a solution with our problem. Thinking is simply surveying our downloaded memory of sensory events rather than reality. The memory contains a huge number of images, patterns, and associations mixed together in a chaotic mess. To think is to send some electric current into the chaos of memory on a fishing trip. When the current passes through a pattern that we can associate with our problem, we focus on it. We then say that we have an idea. Ideas are very much like sub-atomic particles in that all different forms already exist as potentials. The one form that materializes as your idea is simply the one you focused attention on (energized).

The analog memory is a mindful memory which means that it is not located or constrained to the lower dimensional physical structure of the brain or body tissues. It is a part of the implicated universal consciousness. Rupert Sheldrake calls it the "morphogenic field". *VII 15 It contains a great deal more data than the physically located binomial memory does. Binomial memory only includes the limited sensory organ data. That data was also filtered through a censoring process. Therefore, analog memory contains some patterns and relationships that never got into binomial's limited memory bank. These patterns or potential ideas are the source of many of our so called original ideas, intuitive insights, and creativity. The savant's exceptional abilities are derived directly from patterns stored in this analog memory. Some of these patterns reveal universal truths or laws that manifest only in implicated dimensions of reality, and these may be beyond our physical abilities to ever perceive, experience, store, and or understand.

For most of us, going to the store for some milk is a task that can become routine. Because it is familiar and boring, the binomial mind, which is in charge of this type of activity, tries to download the process into the subconscious as a conditioned response. On the way to the store, we will be engaged in thinking about past experiences or future plans while the monotony of the present tense situation is largely ignored.

I often arrive at the store without remembering the trip at all. This is exactly what we do with language. We go from raw data to conclusion without being aware of the process that got us there. The vehicle of language is simply taken for granted.

I could select to go to the store in a car, on horseback, in a wheelchair, on a bicycle, or by foot. The experience in each case is entirely different. If I decide to walk and it starts to rain, I probably will decide not to go at all. If I decide to drive, I will go all the way into town and mail a letter while I am at it. The vehicle changes the experience, and determines the outcome, in many cases.

We can go from raw data to conclusion with English, Hopi, or Chinese. The experience in each case is quite different and the conclusions are often at odds.

Thinking is talking to ourselves and we use our native tongue to do that. The language we use determines our thoughts. If your language is English, German, or Spanish, you will automatically tend to divide the world up binomially and process in terms of cause and effect. If you normally speak Hopi, you will perceive a more integrated analog world and process the data in terms of relativity.

In the nineteen thirties, Benjamin Lee Whorf coined the term

"Linguistic Relativity" to describe this effect. Linguistic relativity = All observers are not led by the same physical evidence to the same picture of the universe, unless their linguistic backgrounds are similar. He also said, "We dissect nature along lines laid down by our native language," and "The grammar of each language is not merely a reproducing instrument for voicing ideas but rather is itself the shaper of those ideas." *VII-1

For different cultures using different language bases, the world is not the same place with different labels as interpreters might lead us to believe. The world is in fact a different place. A checker player does not perceive or think about the red squares.

Each language is an informal logic form. It has a premise that is expressed in that culture's worldview, and the rules of processing are outlined in the syntax of that grammar. Most languages are swollen with perturbant mechanisms so that they can deal with the full scope of reality.

Cultures of the western world developed languages based upon abstracted words. English, Spanish, and German chop the world up into separate definitive parts, which is the definition of the binomial mode. Many aspects of these cultures reflect that basic orientation or perspective. Important functions within these cultures are institutionalized. We see separation of church and state and occupation from family life, for example. God is perceived as a separate entity located somewhere else like Heaven. Is there any wonder that God's gender is always masculine in these cultures? These languages form sentences that are an expression of cause and effect. The sentence structure and the reasoning that results are vectorized or linear. Noun → verbs → object. Rational thought and reasoning is the preferred way to deal with life's problems. There is a perceived need for a consensus of opinion because there can only be one correct position. There is a general sense of the vectored flow of time and great value is placed on moving on, growth, progress, change, and improvements. Achievement or goals are the reason for present tense activities and children are of primary importance. Youth is the preferred state of being. The vector is the symbol for the male gender. Matter and the nouns that represent it are the primary focus. Children learning to speak in western societies are encouraged to start with nouns: mommy, daddy, horsy, and doggy. (An interesting note here: We intuitively put the "y" on the end of these words to soften the materialistic idea and make them more like verbs or processes.) There is a tendency to materialize reality (Reify) by using nouns to represent processes and other non-material phenomena. "July is hot," and "Diseases cause symptoms," for example.

The verbs of being are the perturbant mechanism used in English to deal with non-causal phenomena. They refer to the things that just are. The verbs of being can not really be defined using the cause and effect syntax of the language. When westerners attempt to communicate with people from other cultures, we intuitively know that translating the verbs of being is impossible, so we simply leave them out. We say "Tonto good Indian."

Newtonian physics was developed by these western cultures because they have a binomial linear type of language form and subsequent world view.

Many cultures around the world developed more analog forms of communication, and communicate basically and primarily with pictures. The Amerindians, Australian Aborigines, and the Chinese are some good examples. The logic of a language based upon analog pictures graphs out as a circle. Consequently they are programmed to perceive circles in the background chaos of reality rather than linear vectors. Circles take on primary significance in these cultures' philosophies and strategies. The Amerindian teepee, kiva, circle of nations, dances, the four directions ritual, and the medicine wheel all reveal this basic orientation. The Chinese duo symbol, cycle of elements, the circles of support and control, and the circulation of Chi are also good examples that reveal this perspective. The big difference in thinking can be seen by comparing military strategies. Amerindians in the old west typically circled the wagon train or attacked individually, while the Federal troops normally deployed collectively in straight lines, columns, and vectored charges. The female, which is the more analog gender, is symbolized universally as a circle. Because analog logic does not abstract reality, life is perceived in terms of wholeness or inclusion. Religion, occupation, language, and life style are one and the same thing. God in analog cultures exists in and pervades the real world. The Great Spirit of the Amerindian traditions lives within the rocks, trees, and the body of man. That spirit is genderless and the female is generally believed to provide the best connection to it: Mother Earth and White Buffalo Woman. The Catholic religion in the west is the obvious exception to this comparison. The Hopi language has no explicit or implicit reference to time. These cultures see the present in terms of the past, and revere their ancestors and their ways. Rituals connect them to the past way of being and are of primary significance in their cultures way of life. Analog information obtained from dreams, vision quests, and dowsing is often valued more than rational binomial conclusions. They appreciate the individuality of enlightenment and feel no need for a consensus of ideas and opinions. Each shaman has his own personal technology and rituals for example. They generally accept the position of others and are not inclined to proselytize or argue their point of view because philosophical agreements do not prove anything to them. Verbs are the focus in these world views and are typically the first words learned by children. Many realities we think of as nouns are verbs in these languages: Like cloud, July, Hopi, and Chi.

Because the Hopi have no concept of flowing time, they have no idea of simultaneous events. Therefore, their grammar contains the built in assumption of relativity. For people who normally speak and think in English or German, it takes a genius to come up with the theory of relativity. Every Hopi child, on the other hand, assumes and expresses that theory every time he or she thinks and speaks.

THE ANIMAL CONNECTION

In a wonderful little book entitled <u>King Solomon's Ring</u>, Konrad Lorenz pointed out that we could, in fact, talk to the animals like King Solomon or Doctor Doolittle. The trick was, animals didn't, and in fact, could not speak English. It was up to us, as intelligent beings, to learn their language. *VII-2 And as Temple Granden will testify, their language is primarily analog mode and pictographic. *V-2

The original horse whisperer, Tom Dorrance revolutionized horse handling in the western world. He is undoubtedly one of the greatest animal communicators of all time. I have studied his approach in some detail, and it is definitely analog. Because he developed his techniques intuitively and not rationally, he had considerable trouble communicating his ideas and method to others with binomial language. First of all, he individualized every training session by establishing a two way communication with the animal. He listened to the horse before deciding what to do. He then used that information to set up the situation so that the horse could eventually discover the correct thing to do. The right behaviors were positively rewarded and negative behaviors were purposely associated with aggravations. Rather than doing something to the horse, he simply directed the animal's own efforts into compliance by way of communication. His system has been spectacularly successful in replacing the older abstracted, standardized, and physical, "doing to" methods that have been popular in this country since colonial times. Horsemen all over the world have adopted his approach because it is simple and it works. Horses <u>trained</u> in this way are much better animals than the ones <u>broken</u> in the older binomial mode traditions. I use Tom's basic approach in analog medicine.

The American Indian is said to have had a similar rapport with their animals. This is undoubtedly because of their basic analog world view. *VII-4 Regardless of language, females, as a general rule, are more analog by nature and find that work and communication with animals comes naturally for them. Males, on the other hand, usually prefer working with machines because they can be successfully manipulated in binomial mode.

The Oriental cultures have analog based languages and so naturally developed their medical and fighting technologies in that mode. Acupuncture, Qi Gong, Tia Chi, and most of the oriental martial arts are solidly based in this mode. In these cultures, both the martial artist and medical doctor begin with the concept of centering mentally in time and physically in space. The mind must stay focused in analog mode present tense where the sensory input can be maintained at all times. They also learn to avoid all binomial judgments which will automatically turn off the sensory input, and trigger typical binomial conditioned reflexes. Both acupuncturist and martial artist attempt to manipulate the energy of their patient or opponent. The acupuncturist redirects his patient's blocked or imbalanced energy flows back into a healthy balance. The martial artist redirects his opponent's aggressive energy back against themselves just as Tom does with horses. *V-3

SUMMARY

We see that what is generally called "thinking" is a complicated multidimensional process, just like all the other aspects of life. The higher dimensional intent of the mind is projected down into all the lower dimensions where it manifests as different functions and things. Science has come to understand that the whole life process is intimately involved, on some level, in a comprehensive communication system. The whole body is part of the "thinking" process, in other words. We do have gut reactions, heartaches, and pains in the neck. The scientific investigation of this phenomena can get impossibly complex and paradoxical if the different dimensional manifestations of thinking are mixed together indiscriminately in a logical treatment. Realizing that this is a top-down process helps us to understand it.

The language we speak determines, to a large extent, how we first perceive reality and then how we decide to deal with it. Western cultures, speaking binomial languages, logically develop binomial linear technologies, and Newtonian physics is the formalized logical expression of that world view. Eastern and Amerindian cultures, using analog based languages, logically develop analog, circular technologies. Relativity and Quantum mechanics best describe the world view and logic of these cultures. Theoretically, the world views' of an Oriental female and a Western male would be the extreme positions in the relative spectrum of perspectives.

HEALTH and HEALING

The failure of objective western medicine to define health and explain healing is perhaps its greatest weakness. Constrained by their decision to confine themselves exclusively to the binomial logic of Newtonian physics, they simply can't define or explain these ideas in objective terms. This supports my thesis that healing and health originate in the implicated dimensions of reality. They must be defined in terms of energy and explained using a formal logic like Quantum mechanics or Super-string theory.

The only definition that is offered in objective terms is a negative one. Health is the lack of any measurable or perceived symptom. Soundness, a word often equated with health, also has a negative definition. It is defined as the freedom from defect, damage, or decay. Wholesomeness, often substituted for the word health, is defined as promoting or conducive to good health, and that is a circular definition. Well-being is a term often included in definitions of health. This is a solid acknowledgement that health is not of physical objective reality but of energetic subjective being.

Without a positive objective definition to refer to, the binomial scientists must logically focus on some measurable qualities. They can quantitatively measure different physiologic parameters and compare these values to those of the general population. A measured difference is considered significant. Symptom is another word for this. A list of symptoms defines a disease or condition. Instead of focusing on health, they zero in on the measured deviations from normal or average parameters and the disease conditions defined by them. No one in this camp is focused on health and healing, because they simply cannot define or measure them directly.

When attention is focused on the disease condition, binomial logic tends to abstract it as a noun. In the syntax of western grammar, nouns do or are things. In other words, we materialize (reify) them. Diabetes causes cataracts and hypertension can cause strokes. As nouns the disease can mentally be separated from the patient, analyzed and treated as such. Western medicine diagnoses and treats colic, cancer, diabetes, the flu, or a wound. Western medicine can have specialists such as oncologists, who specialize in diagnosing and treating cancer. The patient is just background, or substrate, for their actual interest and focus, which is the measured deviations from normal.

A noun is also perceived as a separate external entity, an entity that can invade or attach the patient. The typical binomial response is to negate, destroy, or eliminate the perceived invader and to repair or negate the damage. The orientation is to employ external energy in the form of physical force or materials. Diseases and symptoms are negated and eliminated. They have antibiotics, antihistamines, painkillers, anti-immune system drugs (cortisones), corrective surgery or physical adjustments, all of which are called the cures for the disease. A cure is generally a noun and comes from outside. But it can also be used as a verb as in "He was cured." Healing, on the other hand, is always a verb.

The story of King Arthur's quest for the Holy Grail describes this western logical orientation to seek salvation or a cure in a physical object somewhere out there.

Energy oriented cultures, using analog mode logic, have been able to formulate comprehensive definitions of health. The Chinese, for example, equate health with balance. They talk about balancing the five elements; balancing yin and yang, and balancing the flow of chi.

Their ability to define health allows them to focus on it specifically as a goal.

If health is balanced vital energy, it is a condition of the patient and cannot be abstracted or reified. A balanced vital energy is the same thing as the healthy patient, and focusing on the balance is focusing on the patient. In the energy modalities that use analog logic, we treat patients. Moreover, we treat individualized patients. We cannot conceptualize pigeonholing the individual into a disease category or focus on them as such. The orientation is from the inside, and the focus is on the patient's own energy. The strategy here is to strengthen, support, and assist the patient's implicated plan. A balanced healthy system will resist influences that create disease conditions. The patient learns to heal and be healthy.

The approach is almost always positive and additive. External energy remedies like herbs support the vital energy balance by adding to the deficient energy element or elements. We change undesirable frequency patterns by combining them with other frequencies. We offer alternative strategies to deal with life's challenges in the real world.

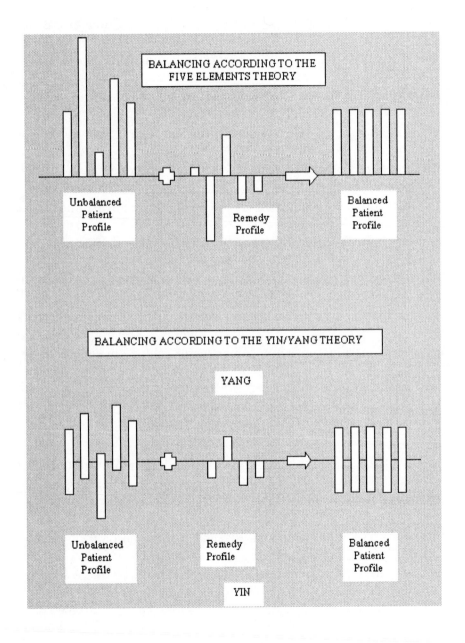

THE MULTIDIMENSIONAL PATIENT

Let's look at the patient from the standpoint of dimensions and quantum reality. Everything is energy to begin with. Some energy manifests as or is frozen into matter that we call biochemicals. Life is not just biochemicals. The chemicals must animate or move. Newton defined energy as the ability to perform work or move matter. Life is more than animated chemicals, isn't it? The biochemicals must move in a very specific and organized way. The code or blueprint that defines and enforces this organized movement is also energy, but it is implicated in the higher dimensions. Hidden like our magnet beneath the paper supporting those iron filings in our analogy in chapter three.

DISEASE

The material body is a physical manifestation of the higher dimensional master plan (Soul) projected down into the lower four dimensions. External forces in physical reality can damage or alter that physical form. Healing the damage caused by these external influences is simply a matter of re-manifesting the master plan. Internal imbalances or blocks in any of the lower dimensions will interfere with and alter the plan's projected manifestation. To re-manifest or heal a defective, diseased physical body, you have to remove the imbalances and blocks from the dimensions above that are distorting its healthy manifestation.

Because the dimensions are nested or inclusive, problems in any one dimension will eventually manifest in all the dimensions below it. Problems that originate because of a poorly evolved spirituality will eventually result in problems in the mental dimension, such as self-destructive or inappropriate logic. This will eventually get you in trouble emotionally, imbalance the vital energy flow, and manifest finally as physical disease. Western medicine recognizes that inappropriate emotional responses like unresolved stress can activate physiologic mechanisms, which in turn interfere with or prevent healing and maintenance of the physical body. Spiritual problems project down and eventually end up as physical disease.

> EXAMPLE: Spiritual dimension = We are different than they are.
> Mental dimension = They are out to get us.
> Emotional dimension = We must be ready to fight or run.
> Energetic dimension = High blood pressure.
> Physical dimensions = Cardiovascular disease.

In such a nested multi-dimensional system the strategy for

successful medical intervention is to identify and focus on the highest dimension involved. Reality and corrective effects project top-down, because each successively lower dimension manifests fewer qualities of reality. Corrective adjustments made on any one dimension can only deal with the qualities of reality manifest at that level or below.

EXAMPLE: Suppose I have a damaged three-dimensional object. My cube of substance has a corner broken off. It is defective in size, shape, and weight. If I attempt to repair it on the first dimension, I can only get a scaffold or tinker toy like representation of the original size and shape projected over the missing corner. A second dimensional correction can deal only with the apparent size and shape aspects of the problem. I can only veneer the defective cube in the second dimension. The first and second dimensions do not contain or manifest the all-important element of mass I need to completely re-establish the original three-dimensional structure.

This is the same in medicine. I cannot effectively correct a higher dimensional problem in a patient from a lower dimensional perspective. You can not actually correct any emotional, mental, or spiritual problems with strictly material remedies or physical manipulations.   Their physical effects or symptoms can be temporarily masked however.

In practical medical applications, we can tell when our focus is too low. First, the condition relapses after an apparently successful treatment. Second, the solution becomes complicated with perturbant adjustments.   Third, inappropriate and paradoxical results are obtained (negative side effects).

We can resolve problems in the multidimensional patient by working on the level where the problem originates. However it is usually easier and more effective to correct a problem from a dimension above it.

Reprogramming or learning is frequently used to resolve the problems originating in the lower implicated dimensions of mental processing. Talking therapies such as counseling or psychotherapy are commonly employed, but these solutions can take a great deal of time and effort and are often ineffective in cases like substance abuse. A spiritual adjustment often provides a solution in these cases.

Emotional responses are generally subconscious conditioned responses. Automatic triggers rather than rational decisions normally activate them. Inappropriate emotional responses can sometimes be successfully redirected or reprogrammed. The rational approach of deactivating the triggering mechanism with conscious recognition and understanding is usually faster and more effective.

Diseases originating in the fourth dimension (movement) are disruptions in the even flow of energy as it attempts to follow the implicated order. This distortion can be recognized as an energy imbalance or block by clairsentience. Symptomatically these manifest as a loss of function, pain, tension, or just a sick feeling. They can originate from direct exposure to toxic, deficient, or excessive subtle energy sources in food, water, and air. Any subtle energy source such as magnetic fields, environmental bad vibes, radiations, or even sound can influence this over-all balance. Unresolved, these imbalances will eventually manifest as physical problems. Problems originating at this level are often directly resolved using analog logic and energy technologies such as homeopathy, acupuncture, Qi Gong, Prayer, or Reiki. The simplest and best solution however is to just rationally avoid the toxic exposure.

Diseases originating in the physical dimensions, such as infections, poisonings, deficiencies, and trauma, result in the physical loss of function and/or form. These types of problems can be effectively resolved by first using the logic and methods of Newtonian physics which applies to these dimensions. Western medicine excels in the arena of the ER. In these cases, we eliminate the perceived "cause" using binomial linear logic. We set the stage for healing by first physically eliminating etiologic agents, setting the bone, or suturing the wound. Healing can then be reinforced and encouraged by shifting into analog mode and applying one of the energy modalities to the higher implicated dimensions.

The word disease is derived from the prefix *dis* meaning undo, negate, or reverse and the base word *ease* or easy. So it basically means "Less easy." A diseased patient finds it less easy to manifest their master plan. We recognize less easy, or disease, as an increased state of tension in the system as a whole. This tension is created when the reality of the lower dimension is distorted and does not match the organizational blueprint stored in the higher implicate dimensions. Tension can result also when two different physical realities conflict.

A pathologic organism, predator, or assailant has its own master plan to manifest.

ANALOGY: Water flowing in a riverbed is matter following an implicated plan or law. In this case, the law is gravity; "Get to the lowest possible place by the shortest possible route."

If a rock is placed in the riverbed, it makes it "less easy" for the water to follow the plan. Therefore let's call the rock the cause of disease. The water flowing around the rock has to travel further than the rest of the water. This takes more work and creates turbulence and tension in the whole system of the river. Work done by the water either moves the rock or eventually wears it away. The ripple is a symptom, or a measurable deviation, from normal easy flow of the water in the river. A materialist would typically say that the rock causes the disturbance or symptom. But from an energy point of view, the water causes it. A rock in a pond does not cause a ripple. If there is no flow, there is no ripple, and if the flow is intensified, the ripple or symptom will be bigger. Therefore we can say that the symptom is caused by the water's compulsion to follow the implicated law, even in the face of resistance.

Vital energy spent on re-manifesting the original pattern or plan produces what we call symptoms (ripples). Symptoms come from the patient, not the disease. Young vital patients have more energy to spend on the process and often display stronger symptoms. If we reorganize a patient's program to make it more efficient, or if we supply them with more vital energy their symptoms will intensify. This is called a healing crisis.

To reiterate: Disease is a less efficient manifestation of the master plan, characterized by resistance or increased levels of tension within the system as a whole. Pain, for example, is a local manifestation of tension. The healing process is a re-manifestation of the implicated order, and it is characterized by a general over-all reduction of tension. Health is the most energy efficient manifestation of the implicated plan.

A sudden release of tension in the patient means either that an energy block has been successfully circumvented, or that patient has exhausted his reserves and does not have enough vital energy left to invest in the healing process.

THE RELEASE

To evaluate the effectiveness of a particular corrective measure on healing, I focus on over-all tension in the system. My goal in treatment is always to have the patient release the excess tension. Releases indicate healing and they are easily recognized.

Release = healing   No release = no healing

1. An over-all reduction in muscle tension becomes evident. Depending on how much energy is blocked, this can be a subtle posture change or a dramatic collapse. Horses lower their heads and cock a hind foot. Stallions and geldings will drop their penises from their sheaths. In most humans, a shoulder drop is evident. Dogs stop wagging their tails and lie down.

2. A sweeping, sleepy sensation is obvious. This is manifest as yawning and a "Doe-eyed look." In extreme cases, humans may faint. Dogs usually go to sleep.

3. People often cry, laugh or have to urinate suddenly.

4. Horses, people, and dogs all lick and work their mouths.

5. A deep sigh followed by a slower breathing rate is very common.

6. Pain is less obvious or disappears entirely.

7. A deep sense of well being, unity, warmth, trust and love becomes evident. Horses frequently attempt to engage you in reciprocal grooming.

We all love releases and will voluntarily seek them out in life. Healthy individuals will purposely create tension so they can enjoy suddenly releasing it.

1. Many animals will fake a spook and run just for the hell of it.

2. Thrill seekers intentionally put themselves in harm's way.

3. Sexual orgasms are releases after physical stimulation builds up tension.

4. Games create tension by holding off the reward of winning.

5. Problem solving such as puzzles create frustration or mental tension.

6. People enjoy watching a scary movie or listening to a spooky

stories.

7. First thing in the morning we often stretch as far as possible, tensing our muscles to the maximum before suddenly releasing them.

I have equated healing and learning, and you can see that graphically demonstrated in training horses. When you present the horse with a cue, he becomes frustrated trying to decide what to do. His whole body reflects the mental tension he is experiencing. When he finally realizes what you want, he will display a physical response exactly like the one I described above in healing.

A particularly anxious or suspicious animal may initially resist your healing efforts. In those cases, I ask it to do a couple of physical things like backing up or coming to pressure. The releases that he experiences upon the successful completion of these simple physical tasks, establish a communication line between the two of us. He will then be primed to listen to the healing cue and respond with a similar healing release when asked.

NOTE: The sensation of release is generic. Be careful working with breeding stallions. They will often associate the release of healing with that of an orgasm and try to mount you.

SUMMARY

The language and binomial logic of western cultures makes it difficult for them to conceptualize and define implicated energetic principles like health and healing. Without any measurable quality to refer to, they have problems focusing on them as goals in their "Health professions." Chinese, Indian, and Amerindian cultures, however, have been able to specifically define these ideas and consequently focus upon them as goals. The explanations and descriptions of healing and health in these cultures are based upon their analog worldview and are very similar to the formalized quantum logic of Western science. A quantum definition of health would be "The most efficient manifestation in physical reality of the implicated master plan." The definition of disease, on the other hand, would be "a less or dis-easy manifestation." A diseased system or organism must work harder to overcome entropy and is recognized as containing more over-all tension or resistance. A sudden "release" or reduction of tension in the patient's system defines healing and can be used to evaluate treatments.

Energy treatments are more effective and produce more profound results when they are applied to/from a dimension above the one where the imbalance or block originates. For example, spiritual solutions best resolve mental problems, and rational understanding more effectively resolves emotional problems. Physical reality is the result of a top → down dimensional projection. The most effective treatments are also top → down dimensionally.

MOVEMENT - CHANGE - LEARNING - EVOLVING

In this chapter I will achieve the impossible, so pay close attention because it can be difficult. Using a binomial language that includes the cause and effect premise, I will describe a reality that cannot be understood in that context. Everything I say here must be qualified. At best, you may get a sense of it by virtue of analogy.

Analog, implicated reality, is the energy that is potentially everything. It is at once nothing and everything, zero and infinity. It contains or is the unchanging idea, master plan or rules of nature that many refer to as God, the Soul, morphogenic field, and /or universal consciousness. It has a magnetic, quantum field like quality, and is characterized by the right side of our dichotomy list. Analog reality = no thing. The lower dimensional physical reality of things is a part of it. So now you see I have just described nothing, or what the Chinese call Wuji. *IX 7

Physical scientists tell us that creation was an explosion. They are of course referring to the creation or manifestation of the explicated, physical reality of things within the implicated reality of no things, which I did not actually describe in the previous paragraph. Thinking

143

requires a focus or a point in Wuji. A point in nothing is the definition and creation of the binomial mode. We now have either a point or nothing. A point does not have a dimension.*IX-1, 4, 5, 6

Science now acknowledges that reality and our mental representation of it have a basic holographic nature. Information stored in memory, for example, is in holographic form. Holographic pictures are produced by splitting a light beam (energy) and then recombining it on a photographic plate after one half is reflected off of a physical object. The holographic aspect of basic reality is in fact created by first splitting the original universal unity (consciousness) into implicated and explicated parts. The reflected explicated energy creates a holographic interference pattern when its perspective is interfaced with the perspective of the implicated field. In other words, recombining the analog and binomial perspectives creates the holographic three dimensional effect we recognize as physical reality. (see diagram on page 98)

The physical scientist's exploding point separates zero and infinity → Two or more moving points of focus = dimensions → Space-time → Explicated material reality. As energy from this explosion moves away from its initial zero point toward infinity, it passes through the stationary magnetic or quantum field pattern of the analog implicated reality. Modern scientists call the stationary field the "Higgs field" or quantum field. This ultimately results in a holographic interference pattern that is manifest to us as three dimensional material reality. No explosion, no movement, no change, no physical reality such as the patient's body. Gravity, a property equated with the material reality in the third and fourth dimensions, is perceived as weight, mass, and substance. It is, in fact, a manifestation of the movement created by the initial explosion and is proof that it is still going on today. Gravity is more; it is in fact accelerated movement, which means that the path of expansion is always changing or curved. In other words, space is curved. Physical explicated reality ends when that exploding energy eventually spirals down into a black hole where it once again becomes a single point with no dimensions.

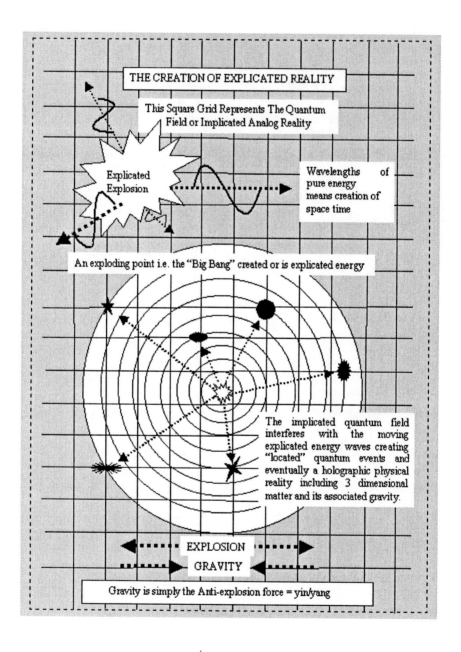

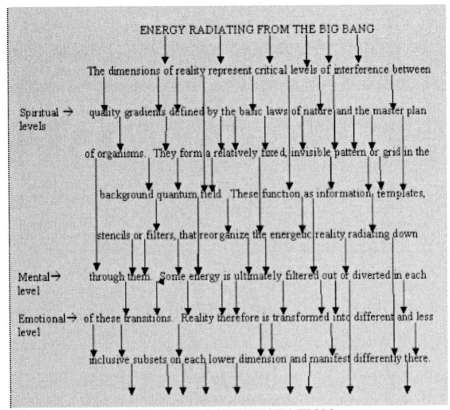

PHYSICAL MANIFESTATION

The Quantum field quality gradient lines are similar to magnetic force field lines. The spaces between the information in this diagram represent the dimensions of reality. The total energy input is divided into different parts on each dimension. For example, the <u>space-time</u> part of the upper dimensions becomes <u>space</u> and <u>time</u> in the fourth. In this analogy, each dimension displays a different combination or pattern of lines projecting through it. In a poor but graphic analogy, these line patterns resemble the informational bar codes commonly used in commerce to identify products.

So what is the significance of this observation? It means that change is a fact of all physical reality including the body. You never step in the same stream twice, and you never take two breaths in the same body or even the same universe. Stability and solidness are illusions created by the limitations of our sensory organs. The enduring changeless quality of the night sky that man has depended upon since the beginning of recorded history for navigation and his calendar of time is, in fact, an illusion. Those heavenly bodies are actually speeding away from us at tremendous speed as the known universe constantly expands from that initial explosion. The body and the physical world around it are more processes than things.

The laws of Thermodynamics clearly articulate this very fact. "Total disorder or entropy always increases with time." Entropy actually is the measure of time. Order was maximized at zero time and entropy maximizes at infinity.

ANALOGY: Within the explicated reality of the material world the atmosphere is basically invisible or implicated <u>like</u>. The atmosphere contains water in the form of an invisible gas. The rotation of the earth and radiation exchanges can result in localized areas of warm air. Warm air has a lower specific gravity than the cold air around it and therefore moves up. This moving gas (draft) passes through an implicated gradient or grid in the relatively stable cooler atmosphere. That gradient is created by the interfacing of the laws of gravity and thermodynamics. A critical point (height) on this gradient is reached where the water in the draft transforms into visible vapor and manifests suddenly as a cloud. That cloud is constantly generated at its lower border as long as the draft continues. It disappears on its upper and leading edges where it is cooled and dissipated by the wind back into an invisible gas. We can see and identify the cloud as a thing because of the relative limits of our perception from our position on the ground. On closer inspection, the cloud is, in fact, a dynamic process. It manifests when drafts of humid, warm air move up through the relatively implicated, stable atmosphere and the frame-work of laws which ultimately define and determine it. The patient's body is like a cloud.

Life is a localized, temporary state of relative order in a generally disordered, chaotic environment. An organism is an open system that takes in order from the surrounding environment and gives off disorder or entropy. The organism uses vital energy to do this work of pumping. A healthy efficient organism spends the minimum amount of its vital energy to accomplish this while a dis-eased or inefficient organism is required to invest a great deal more. A healthy organism, consequently, appears more vital and younger to us because it has more of this vital energy left over to invest in less vital activities.

Evidence of life is universally considered to be the appearance of order in a generally chaotic environment. A living organism is a temporary, localized reversal of the basic law of entropy. Increasing entropy in the universe is observed as an over-all shift to the red end of the light spectrum. Clairvoyants report that the colors of the seven major energy chakras associated with the physical body are in the reverse order. The lowest, root chakra is red, and the highest crown chakra is violet. This seems to be a graphic confirmation of life's anti-entropic quality.

The master plan of life reflects the basic dichotomy of order/chaos created by the big bang. The physical elements created by that event are inexorably associated with space/time and increasing entropy/chaos. The separation of things defines binomial mode and that ultimately translates into more things or complexity over time. The fourth dimension manifests the most individualized physical parts and complexity. These are balanced against the equal and opposite qualities of unity, simplicity and organization of the analog mode. The first and last dimensions manifest unity or the least number of parts and complexity. The life process defined by the master plan seems to be a cyclic progression from unity and organization into complexity and chaos and back again to unity and organization. The fourth and eighth dimensions are the watersheds where the balance shifts. There are three dimensions above the eighth, between the eighth and the fourth, and below the fourth. At those two junctions clairvoyants see the templating or mirror imaging effect. The fourth chakra corresponds to the eighth dimension and the surface of the physical body to the fourth dimension.

We might speculate that the purpose of the physical life process is to learn from complexity and chaos how to be more organized and in tune with the basic laws of nature. This seems to indicate that progress toward analog unity or spiritual enlightenment requires the experience of binomial separation in the physical realm. The Chinese concept of Yin/Yang and Newton's idea of equal and opposite keep showing up. Balance is the lesson.

Single celled organisms can literally live forever reproducing themselves by fission. The evolution of more complicated multicellular organisms, however, required genetic re-combinations that could be tested and proven in the environment. This ultimately led to the invention of both sexual reproduction and death. Progress or evolution of the physical body is achieved by constantly testing new genetic combinations or ideas while at the same time removing old, outdated ones from the breeding population.

The inherited genetic program orchestrates a dynamic exchange of both energy and physical materials with the surrounding environment. The body remains intimately connected to the rest of reality throughout life, constantly exchanging these elements. The atoms that make up a body for example, are completely recycled every four to seven years depending on which expert we ask. This genetic program includes an initial growth period in which the vital energy actually reverses entropy. This is followed by a period of maturation during which the exchange of entropy and order is relatively balanced. Reproduction takes place during this period. During the next aging phase vital energy begins to wane, and the effects of entropy dominate. At death, the vital energy is lost, and the pumping stops. The second law of thermodynamics is then uncontested and the physical body's relative order dissipates back to the level of the surrounding environmental chaos. "From dust to dust." Life is dust + vital energy + a plan. *X-2

The implicated plan for a particular process or organism physically manifests as genetic material. The higher dimensional, implicated plan spirals down into the lower dimensions of reality and is transformed into genetic chemistry within the cell. The chemistry forming around this energetic projection naturally takes the form of a coil or spiral. The physical genetic material is the mechanism the code uses to manifest a physical body, and it controls only the explicated, physical part of the total organism. Many possible genetic messages or instructions are retained as potentials and are never or infrequently activated. Each differentiated cell type contains the full

DNA code but only uses those genes that relate to its specialized function and location. The higher implicated levels of the organism ultimately determine which genes will be activated, in what cells, and when.

The genetic material or messages frequently change or mutate during the individual organism's life due to interactions with physical reality. Cosmic rays and chemicals for example are known to damage them. Other genetic materials from virus or bacteria are left behind following natural exposures. These changes in the genetic material are retained and passed on to progeny, in some cases. These foreign genetic patterns manifest symptomatically as the classic homeopathic "Miasmas." This foreign genetic material is often permanently incorporated and becomes part of the species genotype. An organism's genetic material therefore is a combined record of both its own and its ancestor's physical experiences in life. Geneticists can actually read the history of the species experiences in its genetic profile. It is a chronicle of a constantly changing and evolving strategy of how best to physically manifest the implicated plan of the soul. *XVI-24

The genetics of an organism combine with environmental factors to determine the present tense physical manifestation of the master plan. The master plan itself is implicated and unchanging. Over time it evolves many different physical manifestations to achieve its ultimate purpose. The individual manifestations or bodies are not ultimately very important or predestined to assume any particular physical form. We can use either a fork or chop sticks to eat our meal. The implicated idea of eating is accomplished as well in both cases. The human body could just as easily have evolved with three fingers or wings. The implicated human being is not constrained by or limited to its physical evolving form. The plan can enlist other physical elements to extend its purpose when the biologic form proves inadequate for the task. The use of tools often provides for a degree of flexibility and potential that the biologic element cannot effectively supply. We can put down the hammer and pick up a pen, for example. The plan's tools are an extension of the physical biologic form and evolve over time in lieu of the biologic form. The evolution of culture and technology becomes part of the <u>process</u> of pursuing the implicated master plan.

The western medical profession still conceptualizes the body in terms of the immutable materialism of Newtonian physics. It thinks of the body as an additive collection of separate functional parts and systems. These functions can be isolated from the rest of the body and dealt with materialistically in the same way a mechanic fixes a machine. A clogged coronary artery can be reamed out. A failing kidney can be replaced. A diabetic can be given insulin. There is no appreciation for the concept of connectedness/entanglement or the dynamics of the life process as outlined by quantum physicists. In the limited worldview of the materialist, energy only moves matter, flows around it or flows along it like an electric current. It is simply the fuel the material body needs to sustain its physically determined functions It is the material of the body that determines function and therefore ultimately life. Function of parts → Life process.

If life and its physical manifestation are actually processes, they cannot be accurately represented or related in terms of binomial, linear philosophy and logic. A process does not have separate and immutable parts. The perceived parts are constantly morphing. While the second and third parts are being measured, the first is changing or mutating. Therefore any binomial evaluation of the life process will always be fundamentally flawed. That flaw will be relative to the speed of the process.

In contrast, the Chinese concept of vital energy or Chi movement does describe the living organism as a process. This theory describing life, as a contained, circulating energy system or process seems to fit the picture painted by modern scientists. The life energy is not visualized as separated from the material of the body as it is in Newtonian logic. In Chinese or quantum logic the material of the body is one form of the total life energy. A piece of ice floating down a stream in spring is the same as the water it is floating in. It is simply manifesting differently. The flow of the Chi, like the river, includes both the fluid and solid manifestations of the body. Working with a dynamic system of flowing energy or Chi is working with the life process not a thing. Life process → Function of parts.

The theory of yin and yang also describes an entangled, dynamic circulating process. In this concept energy moves continuously between the two extremes of any quality without ever stopping in either extreme or at the balanced point between. Stability or health of the organism is maintained by balancing movements into the yin with equal and opposite movements into the yang. Left is balanced with right, up with down, forward with backward, etc. It describes a balanced gyroscope like effect. *V-3, 4, 5, 6, 7

The treatment of a process is fundamentally different than the treatment of a thing. A broken or defective thing can be physically repaired using Newtonian, binomial logic and external energy inputs. Repair is a physical function of the doctor or mechanic and the patient or machine is not actively involved. A fixed patient/machine is theoretically physically reconstructed to assume the same form as the original model. The original organizational plan or blueprint is maintained in tact. This reconstruction is obviously free to recreate the same problems it did in the past.

Working on or changing the basic plan is the way to correct problems with a process. The diseased or injured body is only the manifestation or result of a defective process or idea. The explicated physical plan is only one of the many possible plans or routes that could be utilized to pursue the implicated master plan's purpose. This is a statement of the quantum principle of "Sum over paths." If your car got stuck on that road before, take another road this time. Learning becomes the central issue in correcting a defective process or healing the body. Learning to use a different route, idea or plan changes the outcome or the manifestation. For healing to become health, the patient must learn and accept a different way of processing or being. They must change their mind and emotions as well as their physical body.

Learning occurs in many different dimensions. It happens on the conscious, rational <u>mental level</u> when the patient changes their mind about a particular logic or strategy they have been using. They can simply decide to stop doing a physical activity that is damaging them, for example.

Learning happens on the <u>emotional level</u> when we disassociate an inappropriate schema that is damaging the body. We do this with psychotherapy or meditation for human patients and with disassociation training in animals. We simply teach the patient an alternative and more acceptable rational response to the habitual ones they have been using.

Ron Hubbard identified irrational associations made by the

unconscious or anesthetized mind (Analog mode) as serious problems that contribute to behavior and sometimes physical health problems. Subconsciously associating the dental technician's perfume with your dentistry anxiety is an example. He called them "engrams," and his solution was also a form of disassociation or learning. *IX -2

Learning takes place at the level of basic physical <u>metabolism</u> also. Vaccinations are a way of teaching the body how to respond to a particular antigenic challenge in the future. In this case, the response manifests as a physical immune response or antibodies.

On the <u>energetic</u> <u>level</u>, Acupuncture and Qi Gong re-direct the flow of chi along the meridians of the body. They do not <u>cause</u> the flow any more than a teacher <u>causes</u> learning in a pupil. These techniques teach the energy body a different and hopefully better way of circulating or balancing the vital energy. Homeopathic remedies work at this level of reality as well. They present an alternative energy pattern, logic or idea to the body that could be used in lieu of the inappropriate strategy, which is resulting in symptoms of disease.

Healing is re-manifesting the implicated master plan in physical reality, after disease or injury alters the original prototype. The healed patient now manifests in a newly evolved and improved physical version based upon what was just learned. It will hopefully be a life process that is more effective at reversing entropy and pursuing the master plan than the last one was. Each disease or problem encountered in life therefore becomes an opportunity to learn, change and evolve into a more efficient life process.

A business organization is not unlike a biological organism. The original implicated business purpose and its master plan eventually project down and manifest in physical reality as employees, buildings, inventory and product. In our society the implicated master plan of a business is usually "to make a profit." The lower implicated business plan describes how this is to be accomplished. When that plan is a good one, the business will be efficient and generate a healthy profit. There will be enough resources generated to fund maintenance, pay for the necessary repairs (healing) and even provide for some investment in expansion (reproduction). On the other hand, if the business begins to loose money (disease) the lower implicated plan must be changed before the resources (vital energies) are exhausted and the business becomes insolvent (dies). The re-organized, re-built company will have the same identity and basic implicated master plan. Its new manifestation is the result of a redesigned lower implicated plan and it will hopefully be a more efficient and profitable organization.

SUMMARY

Physical reality, which includes the body, is in fact a process. Simply repairing a physical body essentially takes us back in time, reversing the process, and violating the basic laws of nature. Healing the body process, on the other hand, involves learning, change and progress, which is an expression of those laws.

The process we identify, as the physical body is an integral part of total life process. Once separated from the animating and organizing energy of life the physical body is constrained to follow the law of increasing entropy or process into chaos.

A rational, logical (intelligent) approach to dealing with an energetic process such as life is to adopt a format such as Quantum Mechanics that begins with an energetic premise, but conventional Western medicine limits itself to the exclusive use of Newtonian logic. Its premise, divides reality into separate and immutable matter and energy, and that does not accurately describe the dynamics of the life process. Their logical conclusions are therefore often at odds with the facts (paradoxical). These inaccurate conclusions frequently lead to complex, convoluted, and inappropriate solutions in conventional western medicine.

THE LIMITS OF BINOMIAL LOGIC

The binomial mode begins with the perception of discernable breaks in the even flow of incoming sensory data. Digital is the word used most commonly today for this mode. These breaks define separate units or events that must then be related logically to each other. In a dynamic system, one of the most obvious relationships is an ordering of the events in time. A reoccurring series of events in time creates the concept of "cause and effect." This idea is visualized and abstracted as a linear vector: cause → effect. Binomial mode's attempt to predict the future and understand reality is therefore called <u>linear</u> logic. Linear can be straight lines, curves, coils, or even spiral vectors. When linear patterns are not perceived in a particular system, we say that it is turbulent or chaotic and logically unpredictable. When we assume the binomial mode, we are primed to perceive or abstract these linear series of events from the general background chaos. These capture our attention and the rest of the chaotic data is ignored. The binomial mode, in other words, censors or <u>reduces</u> complicated states of reality to a few perceived <u>linear</u> series. We call them the basic laws of nature. The scientist's job is to reduce the chaos and complexity of reality by identifying and illuminating the laws of nature.

Binomial logic can track an almost infinite number of static singularities, but that tracking depends upon the permanence or immutability of the last singularity measured. For example, if we begin with the series 1, 2, 3, 4 and the first element disappears, we end up with 2, 3, 4. So now is the 2 actually a 1 or is it still a 2? In explicated physical reality, immutability is usually a reasonable assumption. However, even there, in some cases this assumption proves false. Have you ever tried to count fireflies on a warm summer night? Is each blink of light from the same fly moving around or several different flies?

Newtonian physics is the comprehensive science built upon binomial logic that can effectively predict the outcome of circumstances in the lower three dimensions or the real world, if it can be reduced to linear sequences. It begins with the premise that the initial qualities measured are immutable and unchanging. It is limited to the quadrants of three-dimensional reality and cannot relate more than three variables at a time. Newtonian logic fails to accurately predict, and is of little or no use when the circumstances cannot be reduced to three or less immutable linear qualities.

The definition of science is: the logical investigation of reality. The purpose of science is to reduce the overwhelming complexity of reality to a few simple and basic relationships. Focusing on those specific relationships orders the chaos, allowing us to predict. A good and valuable science simply and accurately predicts.

Collecting any and all trial and error data without a prejudice is technology, not science. Potentially a complete trial and error data base would accurately reproduce all the original complexity and chaos of reality. We would end up right back where we started, confused by the complexity and chaos of our own abstract reality. Many fields of study that like to call themselves scientific are actually not. These disciplines are typically drowning in their technological data base and are unable to accurately predict and therefore effectively manipulate their particular reality.

The problem is not simply a matter of semantics; it has some profound and practical ramifications for mankind. A field of study that does not constrain itself with the most advanced and inclusive discipline of logic available will inevitably be over run by the advanced technologies that logic has produced. Using a technology before you have the intellectual (logic or rational) capacity to understand it can lead to disaster. (A child playing with matches comes to mind). Unrestrained technology produces Frankenstein monsters.

A field of study that limits itself, like western medicine has, to the exclusive use of Newtonian physics and its binomial linear logic, severely limits its potential. They will not be able to accurately and consistently predict results or efficiently and effectively manipulate their reality. The following are some technologies that I see currently masquerading as sciences.

METEOROLOGY: Meteorologists have tried for years to use Newtonian binomial logic to predict the weather. Their efforts are domed to failure because there are far too many variables. Computers can extend the logic with serially applied perturbation adjustments to some extent. However almost all the data to be analyzed is mutable: even the observable clouds and storms are actually processes. The atmosphere displays turbulent, chaotic activity that cannot be reduced to simple linear relationships. The weatherman simply predicts that the atmospheric patterns will play out as they have done in the past. They project from an extensive present tense database. When viewed from a total global perspective, however, the weather has probably never completely reproduced the same pattern. The results speak for themselves.

INTERNAL MEDICINE: If Newtonian binomial logic cannot handle the chaos of the weather, it certainly cannot handle living systems either. The animated dynamic state of the living organism has far too many variables, too many dimensions, and the individual parts are also mutable. The individual chemicals of life exist only as an artifact of linear laboratory chemical analysis. In the reality of life, the chemicals are connected in a dynamic analog like soup. The chemicals lose their individual identity, combining many times and in infinite ways to form the physical fluids, hormones, cells, and tissues of the body. The chemical processes in the living body are more chaotic than linear. In Newton's third and fourth dimensional world of machines, the immutable physical parts equal the machine. Life, however, is far more complicated, being made up of mutable physical parts, and several additional implicated dimensions. There is more to life than the sum of its physical parts. The life processes lend themselves more to analog logics like Quantum mechanics.

The FDA, several years ago, banned the marketing of combination products. A number of time-tested formulas such as Vitamins A-D-E and "Combiotic" (penicillin-streptomycin) were pulled from the market. Their scientists correctly based this decision upon the fact that the linear binomial logic they used could not possibly keep track of more than two variables in a dynamic chaotic system, even with the ultimate binomial appliance, the digital computer.

Binomial linear logic is inadequate to deal with complex biological problems. It cannot predict that a potent insecticide like DDT would cause breast cancer or interfere with the reproduction of raptors. Who would have guessed? There is no way that these particular consequences could have been anticipated and researched ahead of time using this logical format. There is no way to anticipate the consequences of a new drug, hormone, or genetic manipulation other than to resort to trial and error or a kind of chemical Russian roulette. Go ahead and try it for a while and see if anything bad happens, has been their approach. The FDA cannot possibly insure the safety of these new products ahead of time any more than they can track products with multiple ingredients with the binomial linear logic of Newtonian physics.

The paradox arises: The FDA, by their own admission is unable to predict the compounded effects of products with multiple ingredients in a living system when they focus on the one thing it is supposed to do. Yet they believe they can predict, with a degree of certainty, the safety of all the many other things it could also possibly do.

The recent problems associated with "approved" products like Fen-Phen, Thalidomide, BST, Baycol, Serzone, Redux and Meridia are evidence of this fact. Believe me, they are only the tip of the iceberg.

One possible way around this dilemma is for the FDA and other regulatory agencies to adopt a more inclusive form of analog logic such as Quantum Mechanics. Another option is to safety test all products both physically (chemically) and energetically. The energetic testing could be easily handled by any number of competent clairsentients. Radionic devises and muscle testing are two obvious alternative protocols that might also be adopted for this purpose.

NUTRITION: We can tear apart dead plant and animal material (food). We can identify different parts/chemicals, and predict fairly accurately how these individual parts will react with each other in a beaker, one on one binomially. Logic leads to prediction and that is science. Biochemistry is a science. But when these same chemicals are added to/eaten by a dynamic animated system like a living organism, there are far too many variables and too much mutability to be analyzed binomially.

Bread historically has been called "The staff of life". Binomial minded scientists decided some time ago that the active ingredient in grain was carbohydrates. Whole grains were therefore improved by "Refining" or "Purifying" them down to that one element. The resulting "White Bread" was considered to be one of the marvels of our modern technological age and even became a symbol for it. Over time research pointed out the need for vitamins and minerals in our diet and so refined sources of these were added to the mix to "Fortify" it twelve ways. Apparently it did not occur to anyone at the time that they were simply putting back in what they had refined out.

Subsequent research disclosed the critical health benefits of fiber and bulk in the diet. Many put two and two together and realized that the staff of life actually referred to the fiber and bulk of the original whole grain breads.

We now know also that the refined white flour is too easily and rapidly digested and assimilated. The consequent spiking of blood glucose levels is directly associated with cholesterol deposition and diabetes. The processing, that we were initially so proud of, turned the staff of life into a high fat, low fiber, junk food of negative nutritional value. Combined with other refined products like white sugar it became a major health risk in the United States. In regard to the subjects of flour and sugar, the "science" of nutrition, based upon binomial logic, proved to be <u>dead</u> wrong.

It now seems ridiculously obvious that our digestive system would be specifically designed to deal with the naturally compounded whole foods we ate for most of our evolutionary history.

Because the nutritionists insist on using only Newtonian physics and its binomial logic, they cannot predict accurately. In just the last few years for example, different authorities in the field of nutrition have suggested several different diets for weight reduction. These authorities all disagree; there is still no science of nutrition.

Medical biologists in the western world cannot logically predict what effect a plant will have on the body after tearing it apart. They practice a type of backwards science that is actually trial and error in disguise. Experience, not logic, indicates that Echinacea is a potent immune stimulant. So after the fact, biologists attempt to explain how just the physical parts or chemistry account for this. They isolate the "active" ingredient/part. The concept of cause and effect is preserved in their eyes, and they ignore the implicated reality of the plant. Then they look for other plants with the same active physical ingredient. They have some element of predictability, but it is not based on logic and science. It is classic trial and error reasoning.

By contrast, in Chinese elemental theory, the whole plant Echinacea is classified as having primarily the qualities of metal and earth. These elements have long been associated with the defensive *Wei-chi* or the immune system. After first characterizing Echinacea energetically, the Chinese would predict that it would be a useful immune stimulant. Logic leads to predictability and therefore, in this case, the Chinese system is more scientific. *XVI-5, 6, 7

PHARMACOLOGY: Years ago biochemists isolated some proteins they called hormones from the chemicals of life. These were produced by the endocrine glands and controlled metabolism. The medical profession was beside itself with enthusiasm, for they believed that science had delivered the ultimate biological control mechanism into their hands. There were hormones for almost everything biological. Estrogen, for example, caused females to come into estrus. Thyroxin controlled the rate of metabolism. A hormone called oxytocin caused milk let down in the mammary gland.

With a full set of hormones, they believed they could theoretically orchestrate the life process at will. This belief was based upon the naïve assumption that each hormone had only one function. They knew that function and named the hormones accordingly in many cases: (Growth hormone, follicle-stimulating hormone, lactogenic hormone, etc.) Binomial logic can handle one hormone = one function. Unfortunately, continuing research began to uncover a much more complicated picture: an analog picture, in fact.

Estrogen did bring animals into heat, but it was also associated with calcium deposition in the bones. It was implicated in hair growth, fat distribution and immune system functions as well. Not only did the hormones have multiple functions, they had complex inter-relationships and negative feed back mechanisms. For example oxytocin caused the uterus to contract only if it had first been primed with estrogen. Binomial logic cannot handle all these variables, so predictability and hopes for biologic control went out the window.

This problem was brought to attention recently when it came to light that young athletes were using anabolic steroids to increase muscle mass and performance. These hormones divert resources from other vital functions to achieve this affect. In young developing individuals, this can short change the heart and reproductive organs. Binomial logic that focuses on only part of the whole picture can result in disaster.

Monsanto Corporation recently revisited this scenario by marketing a hormone to increase milk production in cattle. The idea was that more milk means more profit. This was a variation on the old "Something for nothing," come-on that has always been so successful in the past with flimflam men. Their sales pitch indicated that there were no adverse side effects, which could not be managed. "There is no scientific proof to show that any adverse effect," was their standard retort to criticisms of the product. They effectively turned the tables on the medical wisdom of the day by demanding proof, scientific of course, that it was causing problems. They were never required to prove that it was safe because that would be impossible considering all the variables involved.

In all the debates about this product, no one (except me, of course) ever pointed out that the dairy industry was built upon genetic selection. A successful dairyman in the past was equated with a successful breeder. But if you artificially simulate production, and that results in health problems for the cows, proper genetic selection becomes impossible. You can never know whether the high production, the bad feet, the poor reproductive performance, or lowered calf survival rates were genetic or artifact. In addition, injecting pregnant animals with hormones obviously affects the next generation in-utero. Are the characteristics of next year's calves' incidental or due to reproducible genetics? No genetic selection or, worse, false genetic selection ultimately destroys decades of record keeping and work. The wide spread use of such a product would eventually result in a population of dairy animals that required these injections to produce milk. Would that be an adverse affect for the dairy industry and a benefit for Monsanto?

When we apply an inappropriate logic form, we get answers that are inappropriate complicated and paradoxical.

GENETIC ENGINEERING: This same kind of scenario is being played out today in the field of genetics. Technicians in this field are making wild predictions for the use of genetic engineering, believing they have finally delivered the ultimate biologic control mechanism into our hands. All we have to do is add a gene for the quality that we want and remove genes for the qualities we don't want.

Here again we have people using exclusively binomial logic attempting to reduce the complexity of an analog reality to a simple double variable. In fact genes do not binomially code for qualities. Genes code for specific proteins that may result in particular qualities after going through the complicated analog process of metabolism. It is again a naïve idea that the coded protein has a single measurable function. Geneticists have long ago noted that the qualities they focus on are always associated with many that they are not focused on. For example, cattle have a poled gene that results in hornless animals. Why not breed all our cattle to be poled? Poled Holsteins are always also poor producers. *X-2

This is simply a case of some very accomplished technicians doing something because they can, and because there is money involved. Technicians masquerading as scientists can, have, and will produce Frankenstein monster like results.

NEUROPHYSIOLOGY: I just finished a book on this subject and was quite amused to see how far the "scientists" in this field have come without even confining themselves to binomial logic. In one chapter, the author states that there is no consensus on the definition of emotion or a way to directly measure and quantitize it. He then goes on to classify the different types of emotions that are commonly recognized. In a subsequent chapter, he identifies their location in the brain and how they are connected to other functions and qualities that cannot be exactly defined or measured either. *V-10

This reminded me of once upon a long time ago in a stone-age culture far away when man first discovered how to use abstract symbols and words. It was a magical time when things first acquired their names. A name was the thing, and it could not be changed. The word could be used in communication instead of the thing, and they could be used magically in chants, prayers and hexes because they had the same qualities as the named. The word keeper was an important man in the tribe, because the Gods told him which words went with which things. Over the years, the word keeper was kept busy naming every thing they could find. Then one day every thing was named, and the tribe no longer needed the services of the word keeper. But there was a big problem according to him. There were words left over. For example, what exactly was an "onald?" If the Gods had a word, there must be a thing. So the word keeper was kept busy looking for the missing things that go with those extra words. The word keeper's craft lives on today in fields like neurophysiology.

The binomial logic of Newtonian physics is severely limited in what it can accomplish as a scientific tool. There are also many disciplines today that are falsely passing themselves off as science. The people working in neurophysiology are obviously attempting to explain the brain's function in strictly explicated mechanistic terms. This would necessitate a binomial Newtonian approach, and that means the individual elements studied must be defined in terms of measurable qualities. If you cannot define emotion in terms of a definitive measure, you cannot relate it logically in a system like Newtonian physics that is based on an immutable, explicated, or measurable reality. No measure = no logic = no science.

SUMMARY

In the western medical world today there are some very well established scientific disciplines that are based upon the limited, binomial linear logic of Newtonian physics. Anatomy, biochemistry, and pathology are a few that come to mind. Many other subjects taught in medical school (like nutrition, internal medicine, and neurophysiology) are not scientific, because there is no way to apply binomial logic in these areas where there are so many variables. Since they reject all other logical formats, they have no ability to accurately predict results. The people working in these areas want to consider themselves to be scientists, but they are in fact simply technicians, collecting and cataloging endless reams of trial and error data. Like the meteorologists, they can project this data into the future with binomial computers to achieve some limited degree of predictability. But in the end no logic = no science = no consistent reliable prediction.

We have no choice in this mater. Advanced forms of logic have already produced technologies that the eightteenth century Newtonian format can not technically handle. Manipulating reality with tools and technology we do not logically understand can ultimately lead to disaster.

JUST PRETEND

Our binomial mode controls how we invest our finite psychic energy. He is the administrator of our daily (wakeful) activities. He is the ever-present protector and is focused by way of the sensory organs on the lower explicated dimensions. His limited view of reality leads him to perceive reoccurring patterns like A-B-C and 1-2-3. The conclusion is that A causes C by way of B. The idea of cause and effect allows and, in fact, compels him to predict and manipulate the present situation to produce desirable future events. Prediction is the first step in protecting us from the slings and arrows of outrageous fortune. By focusing on only one particular A-B-C series, he can ignore all the rest of the complicated and consequently confusing sensory data. He is constrained to binomial linear logic, which means that one thing leads to another. With an appropriate plan in brain, he can then decide where and how to invest the finite energy.

Like all administrators, he is very much concerned with efficiency. Wasting vital energy is a cardinal sin. After all, he may need it to fuel a mad dash or fight for life at any time. He is on guard at all times, overseeing and judging how analog mode is spending their valuable resources. When no overt action is needed, he will allow analog to use energy for maintenance and other reasonable activities. "Seriousness" is the word that embodies the essence of binomial mode. It includes both the ideas of cause and effect and judgment, both of which are in the left column of our dichotomy list. Serious consequences are to be avoided at all costs. This is binomial's philosophy and purpose in life in a nutshell. The value of binomial mode's veto power is that it keeps their feet planted firmly in physical reality where their physical bodies must survive. After all, there are tigers out there. The problem with it is that it prevents us from exploring and working with implicated reality when we need or want to. You can't see auras, feel energy or project prayers to others because it just doesn't make any <u>sense</u> to binomial. "Come on, get real." "Be <u>sensible</u>". This is a simple paraphrasing of the binomial qualities I outlined in the chapter on processing.

Beliefs are established when experiences in reality reinforce or prove binomial's mental constructs. They are self-perpetuating and therefore enduring. Beliefs determine to a large extent the consciousness focus. Focus determines what reality or wave function manifests. Manifesting reality in turn substantiates or proves the belief.

Belief → Focus → Reality → Proof

Binomial mode is temporal and therefore perceives and values the idea of progress and the need for change. Binomial mode is innovative. Progress in life is like climbing a ladder. We can safely move up the ladder only one step at a time. The established stability of one foot allows for exploration and the tentative advancement of the other. He has and will adopt new logic forms or beliefs, if they are proven to be more effective. In those cases, where it is deemed safe, he will continue to allow some latitude to explore other options and ideas.

How does a person with a solid foothold in materialistic medicine negotiate the step up into quantum reality and energy medicine? You need to somehow get around binomial's well-established and powerful beliefs without destabilizing him (serious consequences). At the same time, you have to be able to reinforce this Quantum logic with energy perceptions and reality proofs. This is how the new idea becomes an established belief. You need temporarily to suspend binomials controlling and censoring functions to experience this perception. Pretending is one mechanism that achieves this goal.

When we pretend binomial mode is reassured, there will be no serious consequences. Pretending is simply physically acting out a positive affirmation. Research has established that an idea is positively reinforced by the physical responses associated with it, so we are more likely to remember an idea that we vocalize and act out. In neurological terms, we are completing the basic stimulus response loop. What we have here is a loophole in the law which we can exploit like any good lawyer. Assume the playful mindset, and binomial will let you act out almost anything.

As a child, most of us have pretended to be a princess, cowboy or some super hero. The key word here is "be," which means to identify with, in analog mode. If asked, the response was invariably that I <u>was</u> a cowboy. I never said that I was <u>doing</u> cowboy things. We actually became our imaginary hero and possessed all their qualities and abilities. We could leap tall buildings in a single bound. Binomial mode suggested that we were, and analog mode always believes him. Experiments in sports psychology have proven that time spent pretending or imagining perfect performances improve future play more effectively than physical practice. In that case, the analog mode restructures the neuromuscular system to fit the imagined or pretended reality. *XI 6

When you first start to explore quantum reality, let go of your binomial safety net, and just jump in. After all, what harm can it do? We're just pretending, aren't we? If you can't really feel anything when you move a hand over the patient's body or divine a pulse, just pretend that you do and physically act out the process. Observe someone who can do it, and pretend to be them and copy their actions. Say out loud that you can do this, because language is from the binomial mode and analog believes binomial. This is called an affirmation. Remember to frame everything in the positive terms that analog mode can understand. While pretending to feel the energy, let your actions be guided by intuitions. Analog mode is bashful, and if you do not acknowledge its messages by acting on them, it will usually stop sending. Practice this alone or with someone else who believes in and supports the concept. Eventually the feeling will come while you are just playing at it.

One of the biggest stumbling blocks to treating energetically is taking the whole thing too seriously and the fear of failure. Thinking about the consequences of your actions for the patient or yourself puts you solidly in binomial mode (future tense+ judgment). Start doing healing energy work in those cases that are hopeless or where there are no other materialistic options. Try it when you can take the attitude, "What harm can it do anyway?" In other words, you must eliminate the possibility of any serious consequences that will get binomial's attention. This is where the licensed professional has such an advantage over strict energy healers. They can do all that is possible or prudent physically before playing or pretending with a little energy work.

Human patients can also use pretending to facilitate their own healing response in the same way the athletes mentioned above improved their physical performance. Studies of the multiple personality syndromes also reveal how this works. The different personalities sharing the same physical body do not share the same medical profile. If one personality is diabetic, the others may not be. The different personalities are often allergic to different antigens. If we want to control the symptoms of a particular allergy in their body, all we have to do is get them to shift to another personality. Similarly, we can also just pretend or imagine <u>being</u> someone who is healthy. While we pretend and act out that particular scenario, analog mode assumes it to be true and responds accordingly, reshaping the body to fit the new identity. Pretend long enough and it becomes your reality.

This is an example of the quantum principle called "Sum over paths." It says that all possibilities occur simultaneously, in this case disease and health. The possibility that manifests is the one we focused on.

IDENTIFICATION

This brings us to a most important point. When we treat patients in analog mode, the focus is on the individualized patient. That patient cannot be identified with a disease or condition and its projected progression. A balanced patient is in present tense; stop. The physical results or consequences of analog energy work cannot be considered logically. When balance is achieved, you cannot expect a specific physical outcome and evaluate the process accordingly. Successful balancing can result in physical healing, death, or an intensification of symptoms. You can help a patient heal, but you cannot cure his disease in analog mode. They often happen together in reality however.

To say that a child or animal is bad is fundamentally different than saying that they do bad things. If they are bad, we identify them, and there is no obvious logical solution. "Are" is analog (verb of being), and analog is timeless and unchanging. If, on the other hand, we say that they do bad things, the solution becomes obvious in binomial reality. Stop doing bad things. This doesn't seem to be very important at first. As literate rational adults, we are use to compensating for this type of grammatical in-articulation. But younger children, animals, and susceptible analog modes are all unable to negotiate that shift on their own. Being is for life or until a dominant binomial mode suggests an alternate reality.

Identifying patients with a disease condition categorizes them in classic binomial fashion. "This patient has diabetes," is not the same as saying that, "He _is_ a diabetic." No big deal in conventional western medicine, but when you have to deal with your own analog mode that takes everything literally, it is all the difference. A categorized patient comes with a program attached. The doctor of course knows what that is, and expects him to follow it, like all the other diabetics he has treated in the past did. Focus on this probable outcome helps to manifest it in reality. Expect him to die, and he most likely will do it for you. His subconscious analog reactions and responses broadcast this message loud and clear directly to both the patient and owner's analog modes by way of resonance. The doctor's binomial mode judgment is ultimately received as a suggestion by the analog modes of both the owner and the patient and this becomes their reality.

This same scenario takes place with human patients. Patients told that they have X disease, which is always fatal, and they have only so much time to live, will likely die on the date indicated by their trusted binomial mode doctor. The more they know about the disease symptoms, the more they will manifest them. They identify with the disease and live out the textbook prognosis. They can short circuit the process if their own binomial modes reject their doctor's suggestion and successfully replace it with other more optimistic ones.

*XI-1, 2, 3, 4

SUMMARY

The biggest problem in doing healing work is maintaining the analog mode focus. The binomial mode is ever vigilant in the conscious state and is always ready to commandeer the vital energy for doing. Fortunately, a deal was struck long ago that allows the analog to keep the energy when we "Pretend." We can use this mechanism to hold the vital energy in analog mode when we need to.

THE DISIPLINE of HEALING

To heal yourself or help a patient heal, you must access and stay in your higher implicated dimensions by focusing on analog qualities, such as those in the right hand column of the dichotomy list. The healing mode is a non-judgmental, present tense passive focus on that individual patient's being or energy profile. It is focusing from the consciousness of the mind, rather than the senses of the brain.

* XII 2, 4, 8, 13

This can be difficult to achieve in the normal clinic setting. The binomial mode has priority, and it will quickly steal the energy back if you have any kind of binomial thought: like the time schedule, fee, disease, cause, consequences, another patient, your assistant, the breed, etc.

The idea of success is of critical importance to all of us, because it determines to a large extent our future. We tend to repeat activities that we deem successful and avoid those that we have failed at. If we consider ourselves successful, life is fun. Success is defined as reaching or achieving a goal. The truth is, man himself is in control of defining and focusing on specific goals. In fact, they are often abstract concepts which exist only in the mind. We can create our own future by carefully selecting our goals. To become a healer, we need to select some attainable goals.

We all have an individual attention span. This is simply how long we can sustain the tension of a specific mental or physical effort before requiring a release or rest. We need to succeed before our attention span expires or we lose interest, and we need to consider this when selecting our goals. The Chinese are credited with the saying, "A trip of a thousand miles begins with a single step." If we select the whole trip as a goal, the reward may be beyond our attention span, and we end up failing because we do not have the stamina to wait for the ultimate success. However, if your goal is a single step, you can reward yourself many times on the way to the larger goal. You can successfully read a book, a chapter, a page or an idea. You must decide how to frame it for yourself.

Not only do we need to keep our goals attainable, we need to deal with them one at a time. This goes back to the fact that we have a finite amount of psychic energy to work with. This energy fuels life and focus controls where energy is invested. If we focus on the lower explicated dimensions of reality, energy goes into rational thoughts, emotions, and the physical activity that potentially is associated with it. This happens even when we do not actually physically execute the plan. If we focus on historic events like trespasses, accomplishments, failures, or sins, energy is tied up and wasted, because history doesn't change. If we are constantly concerned about consequences and goals, energy is sent off into the future to fuel potential activities that may actually never happen. Energy spent on these past and future binomial activities is not available for physical activity, digestion, and healing in the present tense. Focusing exclusively in the present tense is called *centering*.

Thinking in a vague undefined way is like maintaining two separate thoughts; we have only half as much energy to execute each. With half as much energy to work with, at least twice as much time is needed to accomplish them. Twice as much time may exceed your attention span. The two ideas may also include conflicting mechanisms that cancel each other out. When we focus on one well-defined goal at a time, we become more efficient. Some of us cannot chew gum and walk at the same time.

The total amount of energy invested in a particular endeavor determines how quickly and forcefully it will produce results. If you put all your available energy into one specific task, the mountain <u>can</u> move.

The formula for success is usually considered to be discipline. Discipline is the ability to focus on one specific goal, and the strategy that logically attains it. Discipline also includes an energy investment sufficient to execute the plan and an attention span long enough to accomplish it.

A strategy for successful healing work on yourself or a patient is the same.

1. Adopt an altruistic, nonjudgmental, and loving mindset. Keep in mind that your efforts are simply suggestions and not demands. Research has shown the "thy will be done" prayer to be the most effective. *XVI 15

2. Select one simple, well-defined, and obtainable analog goal, such as the release. Do not go for the whole journey. It usually takes too long to see specific conditions or symptoms relieved.

3. Choose an appropriate analog energy modality that you are comfortable with. I personally use Chinese acupuncture, Qi Gong, and Homeopathy.

4. Assume the "mindful" analog mode. Focus your undivided attention on the patient's energy profile and into executing the treatment plan. Block everything else out, especially the <u>named</u> symptom or condition.

5. Maintain an undivided, intense analog focus until your goal (release) is achieved.

6. Disengage the analog connection and shift back into binomial mode, letting the balance manifest, as it will. Remember you only choose realities; you do not cause them. * XII 14

Get all these ducks in a row and the healing will be quick and dramatic. But let some wonder off, and it can take forever or not happen at all.

MEDITATION

Meditation is how you practice focusing exclusively in analog mode. It is the way you heal yourself, and it is the only way I know to effectively expand your analog attention span. I do not believe you can be an effective healer without meditating on a regular basis.

The ultimate result and goal of regular meditation is recognized as the conscious psychological state we call "**grounding**". The basic brainwave pattern (EEG) of a grounded individual generally displays a lower frequency, over all (calm).

This effect is due primarily to the fact that these individuals have shifted their focused to the higher dimensions of reality. "Be still, and know that I am God;" Psalms 46-10 Rational mental activity is in a higher dimension than the emotional and instinctual. Grounded individuals have successfully reprogrammed their mental activities, deactivating the triggers of automatic conditioned reactions (emotions). Consequently they are more in control of their mental functions. They are able to sustain a prolonged conscious connection (communication) without intermittently disconnecting to emote. The grounded respond (analog) rather than react (binomial).

Many instructors of meditation technique teach that you need to stop moving and thinking. You must do this without falling asleep. Refusing to scratch the itchy nose rejects a physical solution and the binomial mode that controls it. Each rational or emotional thought potentially diverts energy from the healing, analog mode into the doing, binomial mode. This is true, but not thinking can take years of practice, if it is ever really achieved at all. I would qualify these instructions by saying that you specifically need to stop all <u>binomial</u> thought. Visualizations, vocalizations, and thoughts that involve items in the right hand column of the dichotomy list can actually help maintain the analog focus. They can also help to keep you awake. Acceptance, unity, love, connection, and being verbs are all in that list. The subjects of these visualizations or vocalizations should all refer to non-physical or energetic entities such as the patient's subtle energy, angels, spirit guides, or God. The classic chants, intonations, mantras or prayers used in spiritual rituals around the world are excellent examples of these types of facilitators.

Music or sound delivered by way of headphones can be very helpful. The headphones block out extraneous noise from your immediate environment that may distract you. The music should have a basic flowing analog format without lyrics, irregular beat, or any structured repetition that you can recognize and repeat later. Music composed by Mozart is a good example. The music must be psychologically neutral for you, with out any emotional associations. Nature sounds such as rain, forest sounds, ocean waves, and wind are excellent. Gregorian chants, Tibet singing bowls, Amerindian flute and drum, or Australian didgeridoo pieces can also be very effective, depending upon your individual past experiences and associations.

* XII 2, 4, 8

You can also use your meditation time to supplement a patient's treatment, by focusing on them with positive intent. You can visualize stimulating their balancing points or offering them the remedy that you have divined for them. You can also simply say a "thy will be done" prayer for them. *XII-6, XVI-15 Analog mode is fundamentally passive and visualizations that involve you physically doing something compromise it, shifting you slightly back to the left. To avoid this problem, it helps to visualize a spiritual helper, guide, or angel actually performing the task, *in lieu.*

Meditation time is healing time for you. Focusing on our higher, implicated dimensions strengthens our own personal blue print, template or soul, and physical healing automatically follows. With practice, you can learn to focus on higher and higher dimensions. The higher you focus, the more profound the healing. Your brain wave activity changes sequentially as your focus moves upward from the high frequency, low amplitude Beta waves in the lower dimensions of consciousness to Alpha, then Theta, and finally to the low frequency, high amplitude Delta waves experienced in the highest dimensions. *XII-2, 4, 5, 8

Disease can originate in your lower implicated dimensions. Automatic conditioned responses developed early in life or during irrational unconscious states are stored in the emotional dimension. These can be inappropriate and even pathologic in adult life. They are called schemas and engrams. They create energy imbalances by stimulating inappropriate hormone production and stress, which cascades down into the physical dimensions and eventually manifests as diseases like arthritis, heart problems and lupus. We have traditionally treated these schemas and engrams with psychotherapy. A conscious rational understanding of their origin and mechanisms involved usually disarms them, because the mental dimension is above the emotional one and therefore contains, and ultimately controls it. *XII-1, 3

Disease can originate in the mental dimension as well. Mental habits or inaccurate mental maps of reality can lead to inappropriate or compulsive destructive behaviors that ultimately lead to injury and /or disease in the lower physical dimensions. These problems can best be handled with spiritual solutions. Here again we solve a problem by going to a higher dimension.

Programming problems or imbalances in these two lower implicated dimensions, (emotional and mental) distort or partially block the projection of the master plan down into the physical realm. The projection is disordered and that manifests as disease. These diseases can ultimately be resolved by reprogramming these implicated dimensions, and we do that best by focusing on the even higher spiritual dimensions. Meditation or prayer is the traditional way we accomplish that focus.

Emotional schemas, engrams, and compulsive behaviors are all automatic conditioned responses that are activated by triggers rather than rational thought. Schemas are programs or strategies that have been relatively effective in the past and down loaded deeply into the subconscious. The "oral gratification schema" is an example. Engrams are irrational associations implanted accidentally. Compulsive behaviors are what we call habits. To survive intact, vital energy must be constantly invested in these programs to offset the law of increasing entropy that would eventually disassociate them. These inappropriate programs are reinforced and maintained by use even though their results are counter productive. Their triggering mechanisms retain their power of energy requisition by way of belief. For example, some people believe that they cannot relax without a physical facilitator like a cigarette, alcohol, or food. Or they may believe that they cannot possibly succeed. If we focus our attention (a form of vital energy) on the higher spiritual dimensions, the inappropriate emotional schemas, engrams, and mental habits below it are deprived of their vital energy in-put. Without a constant supply of energy in put, these programs automatically begin to disassociate. When these inappropriate implicated programs dissolve, the physical disease associated with them heals as the master plan or template re-asserts itself. It is just that simple.

When the schemas, engrams, and mental habits disassociate, a brief period of chaos ensues before a reprogramming occurs. The chaos creates some mental turmoil that can be manifested as any number of things. While you are in a meditative state, a demon may appear, or an arm or leg may jerk around involuntarily. You may break into a cold sweat, cry, or a feeling of panic may suddenly sweep over you. Any physical, emotional, or mental response (resistance) you initiate in response to these phenomena reinforces the belief in the inappropriate program and infuses it with vital energy. To completely disassociate that program, you must maintain your meditative focus, not moving or responding until the turmoil passes. At that point the schema, engram, or habitual triggers are deactivated. You will still be aware of the triggering circumstances that activated it in the past, but you will no longer be compelled by it. Involuntary emotional responses and addictions become things of the past. With time, they will completely dissipate as new more effective and higher-level programs replace them. The chaos and turmoil of disassociation is exactly the same type of reactions we commonly see portrayed in religious exorcisms. Religious exorcisms often inadvertently direct the focus so high that several inappropriate programs come apart at the same time. The excessive turmoil produced by such an over zealous approach is the subject of legend and several movies. It is advisable and more manageable to disassociate these inappropriate programs one at a time and this is effectively achieved over a period time with a progressive meditation program.

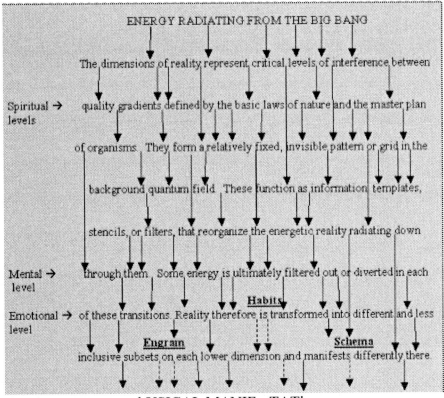

ENERGY RADIATING FROM THE BIG BANG

The dimensions of reality represent critical levels of interference between

Spiritual →
levels

quality gradients defined by the basic laws of nature and the master plan

of organisms. They form a relatively fixed, invisible pattern or grid in the

background quantum field. These function as information templates,

stencils, or filters, that reorganize the energetic reality radiating down

Mental →
level

through them. Some energy is ultimately filtered out or diverted in each

Emotional →
level

Habits

of these transitions. Reality therefore is transformed into different and less

Engram   Schema

inclusive subsets on each lower dimension and manifests differently there.

phYSICAL MANIFesTATion

Emotional schema, engrams, and compulsive patterns of behavior and thinking are all forms of subconscious conditioned responses. They are programs of the physical brain and are activated automatically by triggers rather than the rational process. They are implanted in our lower implicated dimensions and block or distort the proper projection of the higher dimensional, master plan and laws of nature into physical reality. This results in an altered physical manifestation that we recognize as disease.

A physical analogy can be offered from the field of study called *Chaos theory.* Puddles of liquid are organized and held together by surface tension. If energy is added to the system in the form of physical vibrations, the puddle develops a snowflake like pattern of ripples on its surface. The pattern effectively dissipates the potentially disrupting energy input, and the surface tension is still able to maintain the integrity of the system. If the vibrating energy input is gradually increased, the ripple pattern begins to show evidence of structural stress or tension. With increasing levels of energy input, the ripple pattern suddenly disassociates into chaos just before reorganizing in a quantum leap into a new and different pattern. The new, reorganized and more complex ripple pattern is able to adequately handle the higher energy input. Increasing levels of vibrational energy input are dealt with in a series of these quantum reorganizations. A point is eventually reached, however, where the increasing energy input overwhelms the ripple mechanism. The surface tension, at that point, is unable to hold the liquid together, and the puddle disintegrates into chaos. The puddle in this analogy corresponds to the brain, the surface tension to the implicated plan, and the ripple pattern to the schema or engram. Disassociations of an emotional schema or a physical ripple pattern appear to have many similarities. They are both automatically replaced in a quantum leap by a more inclusive, more efficient, and higher level program. The new program, in both cases, can handle higher levels of chaotic energy input. They are driven into higher levels of organization by the force or law of entropy in a yin/yang like mechanism. The force that causes entropy always to increase with time in explicated reality is balanced by an equal and opposite force to be more organized in implicated reality. *IX-4, 5

Practicing some form of energy exercise can be very helpful in maintaining your own health and for expanding your intuitive abilities and insights. Tia Chi, Yoga, and Qi Gong are some good examples. Whichever discipline you chose, practice it in solitude, concentrating on the flow of energy through your body or aura with each movement or position. Develop a sense or feeling of moving through the world's energy field. A basic knowledge of the Chinese meridian or East Indian chakra systems and how the energy normally flows through them will be most useful. Feeling the energy move in your own aura and/or body is the first step to feeling and understanding the process in a patient. *XII-7, 9, 10

Once you become aware of the movement of energy in the body, you can begin to develop the perspective of the "witness," which is very important in healing work. This is the ability to observe the condition of the lower dimensional body, including the brain, from the position and perspective of the higher dimensional <u>mind</u>. Your mind observes the energy profile as an accepting, <u>non-judgmental</u> and <u>unemotional</u> third party. From this perspective you are secure in the knowledge that the implicated, higher dimensional "soul" or master plan is not addicted, in pain or in jeopardy. The diseases of the flesh (physical) are reframed as opportunities for the higher dimensional (emotional- mental) levels to heal, learn, and progress. In chronic disease conditions, the patient's resistance to making positive changes actually creates the tension (stress) in the system. It is as if they are addicted to the particular behavior and thoughts that are creating their disease problem. A healer who empathizes with this resistance to change and evolution will not be of much help in the ultimate healing process. A good teacher challenges and pushes the un-motivated students to move out of their comfort zones of ignorance.

SUMMARY

Healing is a higher dimensional phenomenon, and to effectively influence it, you must learn to assume the perspective of the "witness." This involves working from the perspective of the mind rather than the emotions or rational thoughts of the physical brain. In Buddhist philosophy, this is called "Mindfulness."

Like anything else worthwhile, the ability to manipulate the healing process requires a disciplined investment of some time and effort. Once you understand the basic mechanisms involved, there is a need to practice and perfect the technique. This may involve the formal study of an energy protocol like acupuncture or Homeopathy.

In addition a potential healer must take up the regular practice of a physical exercise program like Yoga, Tia Chi, or Qi Gong, and commitment to meditate or pray daily. The ultimate goal of a discipline like this is called "grounding." The basic brainwave activity of a grounded individual has a lower frequency than most.

The mind is the basic and most important tool of any healer, so you must assume control over yours to be consistently effective in that effort.

The grounded individual is far from an emotionless zombie. He has simply learned to take conscious control of when and how much to emote.

DIVINING

We are working with the higher dimensional reality (subtle energy) of the patient in analog medicines. Energy problems must be treated with energy solutions. The patient's subtle energy profile is balanced by adding a remedy's subtle energy to it or by physically redirecting its flow or distribution. This type of energy is not normally perceived by our sensory organs. So how do you successfully manipulate something you can not physically sense and measure (implicated)? Some times I get help from my psychic or clairvoyant friends. That is the subject of the next chapter. In this chapter I am going to explain how insensitives, such as myself, can learn to effectively access information in those higher dimensions.

The answer is to be found within the basic tenets of Quantum mechanics, which describes reality in those higher implicated dimensions of reality. Connectedness or unity is the best description of implicated reality. Both the patient's and the doctor's higher implicated dimensions are part of a total quantum field of consciousness. Individuality first becomes evident at the implicated level we call the mind. All individual minds are still connected or analog. Analog, then, is both a description of the mind and its mode of function. This is the level of consciousness we have to focus from and the mode we must assume in order to access information about the subtle energy field associated with a patient.

Another characteristic of implicated reality and tenant of Quantum mechanics is <u>negative</u> <u>space</u>-<u>time</u>. There is no quality we can recognize as <u>location</u> in this reality. Consequently the subtle energy of a patient is not confined to the lower physical manifestation's (body's) location. There is also no need, and in fact, no way to actually send and receive subtle energy information without locations.

Reality in the implicated dimensions is totally <u>energetic</u> and subject to the <u>uncertainty</u> principle of Quantum mechanics. This essentially means that the many details we see evident on the physical plain are not obvious here. The mechanisms relating those details, such as cause and effect, are less evident as well. Implicated analog reality is simple. How can the mind of the doctor analyze or influence the mind and subtle energy of a patient? The answer is as simple as analog reality itself: Focused mindful intent.

The <u>uncertainty</u> <u>principle</u> combined with <u>relativity</u> ultimately eliminates the bedrock of physical reality, which is consistency. If our measurement of reality changes it, we can not measure the same event twice. If my individualized mind affects the event, no two observers will see and therefore effect it the same way.

Reality here is present tense only, and individuality rather than categorization is the rule. We cannot rely on memory (past tense) and data collected from others. We can only add a remedy's energy to the patient's and see if it works. In other words, do we get balance and a release? That could involve exposing the patient to large numbers of potential remedies (trial and error).

Fortunately we do not have to be in direct perceptual contact with a patient to pick up or tune in to his energy profile. In quantum reality, the patient's physical location is not relevant; conscious attention is the only connection that maters. We can therefore match our well focused thought or memory of the patient with potential remedies in a controlled setting.

The actual process of direct subtle energy assessment by relative insensitives is called "Divining" or "Dowsing," an activity that many conventionally trained minds have a hard time accepting. It is magical if you insist on analyzing it in binomial mode with Newton's linear logic. However when we analyze it in analog mode using the principles of Quantum mechanics, as outlined in the chapter on logic, it is perfectly logic and therefore scientific.

A quick, and admittedly redundant, review of analog mode functions allows us to appreciate how those qualities combined with the Quantum mechanic principles to explain the perplexing phenomena of divining, dowsing, and shamanism.

According to both Western laterality and Eastern yin/yang theories, every human personality is a composite of two distinct polar entities or personalities. I believe both theories are describing the same fact: the western left is yang, and the right is yin. The left yang is the human doing (physical), and the right yin is the human being (energetic). In the chapter on processing, I discussed how these two modes compete for the vital energy to fuel their particular perspective of reality. The left deals primarily with explicated reality in binomial mode, while the right deals with implicated reality in analog mode.

In analog medicine, we are dealing primarily with the reality according to right analog perspective. Analog mode is in charge of digestion, maintenance, healing, and other functions that do not usually involve voluntary active movements or conscious control. Analog controlled functions are relatively slow and sustained, compared to the binomial ones which are quick and short term.

Analog mode is tuned in to the implicated (subtle) energy information that the physical senses and their neurons fail to transmit to the binomial brain. Analog reality contains far more energy and information than the binomial mode can process.

The physical senses, which are controlled by the binomial mind, provide or define focus for the analog mind. The binomial functions like a radio or television tuner that selectively determines which of the numerous analog frequencies out there we will focus on or tune in to. The radio or television frequency bands are like the different sensory organ's binomial information coming in to the brain. They act as carrier waves for the smaller more subtle information frequencies. My decision and ability to selectively access analog information from that particular plant and not the others growing right next to it, is determined to a large extent by my physical sensory organ focus or tuning.

Analog mind is often erroneously equated with the subconscious mind. The fact is both sides of the brain have conscious and subconscious divisions. The <u>conscious</u> <u>binomial</u> brain plans and executes many of our daily activities. Physical activities controlled by the binomial mode can also be down loaded into the <u>binomial</u> <u>subconscious</u> as automatic, conditioned responses. We are all <u>consciously</u> aware of <u>analog</u> information which becomes manifest to us as "feeling" about something or someone for no apparent rational or measurable reason. The people I call *"sensitives"* consciously access this information, and some may even have problems turning it off. I have found that almost everyone can be a conscious "sensitive" when he or she learn to successfully circumvent their binomial mode's judgment and censoring activities. (See chapter fourteen for more details on this subject). A great deal of the analog information received is stored in <u>analog</u> memory <u>subconsciously</u>. So if you took some time to smell the roses, the energy profile of rose is in your analog subconscious memory bank. This information is constantly used by analog mode in executing its subconscious responsibilities in the body.

Conscious binomial activity can result in damage and imbalances in the physical body. In those cases, analog mode sends a message over to the binomial conscious level. Because analog is illiterate, it must use feelings, symbols, pictures or coded information. Pain, for example, is an analog message to the binomial to cease and desist the particular activity that is damaging the body. Intuitive insights and hunches are other examples. When the binomial shuts down at night during sleep, analog mode is free to communicate more openly with the symbolic stories or messages we call dreams. *V-1

To access this store of analog information consciously, we have to develop a formal communication between our analog mind and binomial brain. This is called "Shamanism," "Divining," or "Dowsing." *XIII-1

Shamans learn how to deliberately turn off the binomial mode and focus or journey in the analog-dreaming mode to access this subtle information. Some of them use hallucinogenic plants for this purpose. These plants, like medical anesthetics, selectively knock out the binomial mode. Accurate interpretation of the dreams can sometimes be a problem, however, because binomial's judgment and censoring functions often influence the translation. Eastern mediation practices can also serve this same purpose. Meditation is the practice of consciously maintaining an analog focus. Thoughts and pictures that come to you during the meditation state can be interpreted in the same way as dreams and shaman journeys. * XVI-13, 14

The different alternative modalities include some time tested and usable physical divining or dowsing techniques that we can adapt for our purpose. Acupuncture pulsing, VAS pulsing, Shu and Ah shi point palpations, radionic stick plates or pendulums, muscle testing, and wand witching are some examples. It is not important for you to be familiar with these techniques at this point. I am sure that you have been exposed, at the least, to the concept of wand witching for water and other things.

These techniques all depend upon a binomial suggestion to the analog that a particular feeling or physiological response, controlled by it, will mean something specific, like yes or no. The value of the information depends upon the specific wording of the question. For example: will this remedy strengthen the vital energy of this patient? Yes or no? How much exactly? One, two, or three units? Analog is obliged to follow binomial suggestions and so this is actually a form of self-hypnotism. *VI-3, 4

Diviners in the past, understandably, have given too much credit to the physical devises they used for this purpose. The witching wand or radionic machine is simply a tool the analog mind uses to organize the communication in a way the binomial brain can understand and accept. The device is just the messenger and translator. Because binomial is almost totally oriented to the physical, it often needs an object to focus on and believe in, this belief can be the source of some problems with consistency. When your attention span is exceeded and boredom ensues, you will try to disengage your focus from the process, thinking that the machine or device is actually responsible for the effect. Turning the process over to the machine or device in automatic mode bypasses the real mechanism of divination, which is your mental focus. Without focused mental attention on the specific question, there is no accurate analog communication, and the answers begin to reflect your personal binomial prejudices. *XIII-1

I am sure that you have caught yourself going through the process of reading a textbook without really getting any meaning from it. To maintain accuracy and retention, you have to match up the effort with your attention span. When your concentration wanes, stop divining and divert your attention to something else. Relax, take a deep breath (center), and begin again.

The pulsing technique used in auricular therapy called the VAS (vascular autonomic signal) is a self-contained diving procedure that requires no device. In this case, analog agrees to strengthen the palpated pulse for yes (positive), and weaken it for no (negative). Any palpatable artery can be used for this purpose. Many use the radial artery in the wrist, but that ties up both hands in the process. The carotid can be used if you want to have one hand free for surveying. You can pulse yourself or the patient while exploring physical remedies, acupuncture points, or diagnostic questions. Muscle strength testing (kineseology) is another self-contained divining procedure that can be adapted for this purpose. *XIII-4, 5

Many people use special music or sounds as meditation aids. These can be extremely helpful for maintaining the analog mode during divining as well. Using ear phones blocks out a lot of extraneous background noise that can distract them back into binomial mode.

I like to divine plant remedies without looking at or reading the name (binomial label) on the specimen tube to avoid binomial prejudices. Later, in binomial mode, it is interesting and educational to try to understand why the remedies supposedly work according to the energy profiles given in a Chinese materia medica, homeopathic texts, or Peter Holmes' books. Plant toxicology textbooks can also give you an idea of how they work from a homeopathic perspective. This binomial intellectual exercise, however, is just for fun. I use the remedies as they are divined regardless of understanding. I usually combine three or four plants and possibly a color in a remedy.

*XVI-5, 6, 7

Analog is also very sensitive and will stop communicating if you question its answers or neglect to act on them. Therefore I always physically move a specimen that produces a positive stick or hit from the general specimen racks to a second open rack. This honors analog's communication, even if I do not actually use it for a treatment. I will usually get several positives moved into this rack, which potentially could be used in a remedy. I then divine the final combination remedy from these select specimens. I have several hundred specimens in my collection, and there are consequently almost an infinite number of possible combinations (of three or four). *XIII-2, 3

I also divine remedies as a shaman in my daily meditations. While floating in analog mode, I visualize walking through the forest or countryside accompanied by the patient. We survey the flora in a generalized nonspecific way, looking at trees, bushes, and plants without specifically identifying any. In many cases, a specific plant suddenly pops into mind, and that invariably proves to be a good remedy for that patient.

I have developed a divining technique that relies on the sound produced by potential remedies. Sound has a very powerful and rapid effect on the energy body. A few seconds of sound from the right remedy is enough to eliminate pain, precipitate an allergic reaction, or affect a release. It is very simple to use, and the mere fact that a person likes or dislikes a sound is information that can help formulate a diagnosis. In other words, we like sounds that tend to balance our energy profile, and we do not like sounds that aggravate an imbalance. See chapter seventeen for more details about this technique.

These techniques involve qualities from the right column of the dichotomy list and deal with analog energy and logic. They do not make rational sense if you analyze them physiologically or mechanically. Analog reality is by definition implicated and materially immeasurable. Belief is the logic that applies here. Attempts to explain the dowsing or divining responses in rational, Newtonian terms is illogical and ultimately unproductive.

In Newtonian logic, we measure twice and cut once. In quantum reality, we have to remember that the act of attention or measurement changes the event (uncertainty principle). It is impossible to establish a second quality of the event or re-measure the first. Therefore, do not question your properly divined answers by rechecking them. It is not only illogical in Quantum Mechanics, but demonstrates disbelief. Remember, judgments push you left into binomial mode.

The analog is a "passive" mode. In analog's implicated reality, where there is no space and time or cause and effect, and therefore information cannot be sent and received. Actively pursuing answers, even mentally, can effectively interrupt the connection by pushing your focus left, into binomial mode. You need to maintain a strictly passive mindset by simply allowing, rather than forcing, the analog (intuitive) communication. *XII-8

SUMMARY

For insensitives to get rational information from our all-knowing analog mode, we have to develop a formalized communication system between our binomial brain and analog mind. Unfortunately the analog is illiterate. The idea behind divining or dowsing is for the binomial mode to suggest to the analog mode a communication code. That code can be based on any physiologic parameter that is ultimately controlled by the analog. One type of response is to mean "yes" and another "no" in regard to specific questions. Muscle strength, electrical conduction of the skin, strength of the pulse, subconscious muscle movements, and symbolic thoughts and pictures have all been used for this purpose. The conscious binomial brain can use this information to formulate a logical treatment plan.

USING SENSITIVES AND SURROGATES

Of all the phenomena concerning energy medicine, the transfer of analog data is probably the most difficult for the Western brain to understand and accept.

Explicated energy is transmitted from one location to another location where it is received. Energy received by sensory organs is transformed into binomial signals by the neurons. It is sent on as electricity in the form of charged ions and neurotransmitters. The brain analyzes this in the binomial mode, the same way a computer does. The process of binomial data analysis we call "thinking." Explicated energy, location, movement, sensory organs, electricity, brain tissue, and computing are all manifest in the lower dimensions. They fit nicely into the cause and effect logical format. This all seems to make perfectly good "sense," doesn't it? [Sensory → Sensible]

In the highest dimensions, energy is implicated and does not move. Therefore it does not have waves, and can not actually be sent. The mind, which is involved with this analog reality, is not located and therefore can not actually receive. It just doesn't make any "sense" when you "think" (compute) about it. I will not actually be able to describe implicated energy in English, as you will undoubtedly notice in the next few paragraphs.

Entanglement is the quantum principle that describes implicated energy: two entities separated yet in touch, what affects one simultaneously affects the other. Implicated energy can best be visualized as an all pervasive web or quantum field. The mind is part of this field and potentially has access to all the holographic information stored in it.

We know from experiments in quantum physics that focused attention is actually a force that can be purposely projected into the implicated field to create interference patterns and subsequently subatomic particles. It functions like the laser beam used to produce three dimensional images from holographic plates. Therefore focused attention is the only mechanism we need to access information stored in the implicated energy field.

At any rate analog information, (often referred to as "subtle energy") is experienced both sensually and extra sensually. It is "carried" by or piggybacks on the large waves that stimulate the sensory organs. It is "received" but not usually perceived by most people. It is downloaded into physical memory along with the sensed data and stored there as a holograph. *III-4

Pure analog subtle energy information also continually "flows" unperceived and unsolicited into the physical body by way of the chakra system.

EXPLICATED-BINOMIAL-BRAIN

IMPLICATED-ANALOG-MIND

When we focus on things we assume the binomial mode and access information by way of explicated energy. Classification and labeling of the thing automatically shuts down all subsequent sensory input. Energy is invested in rationally evaluating the situation and planning an appropriate response. We actively do.

If we focus on implicated energy the analog mode is activated, and sensory input is maintained. Habit and intuition control our response. We passively experience or be.

So we have a problem in treating energetically. We first have to be in analog mode to perceive the problem, but we have to shift into binomial mode to plan and execute a solution. Some people were born with (or have inadvertently developed) the ability to control this shift between modes. With practice the rest of us can learn to logically control the process by selecting options from the appropriate side of the dichotomy list in chapter five.

Healers called "shamans" have learned how to focus on a patient while maintaining an exclusive analog state or mode. This altered state of consciousness is called "Journeying," and during an imagined trip, they perceive diagnostic information about their patient. But even in this dream like state, thoughts about possible physical actions send energy into the muscles that would be potentially used. Simply imagining yourself <u>doing</u> compromises the analog mode, shifting some energy back to the left. Universally, these healers solve this problem by acquiring a spiritual helper whom they then direct to <u>do</u> the actual treatment for them. This allows the shamans to maintain their passive analog state. *XVI-13, XVI-14

People in the west commonly utilize this same mechanism in healing prayers by enlisting God, Jesus, a Saint, or an Angel to do the work. The prayer remains totally passive with their primary doers (hands) folded and their goers (feet) disengaged by kneeling.

*XVI-15, XVIII-1

In the more secular setting of clinical medicine, many of us will have trouble shifting quickly and cleanly enough between these two states to be practical at first. I have this problem myself, so I get help from others to expedite the treatment of a patient. The idea is to have another person access analog mode and stay focused there on the energy. I call this person the "sensitive." I use this term to cover clairvoyants (see), clairsentients (feel), clairaudients (hear), and any other kinds of "clairs" you can think of. They communicate their observations to me, preferably without words. I can then stay in binomial mode first, planning the treatment strategy and then executing it.

The "sensitive" is a person who can sense the analog bioenergy in one form or another. I began with a few select individuals whom I knew had this special ability. I soon found out, however, that almost everyone could do it with a little coaching and suggestion. So now I often use the owner/ handler as the "sensitive." This works even better in many cases, because the animal's problems are often shared with its human. Owners that opt for this kind of therapy are likely to enjoy being involved in the healing process. On many occasions I have utilized groups for this purpose. * IX 7

People differ in how the analog subtle energy manifests for them. For example, some people never consciously feel subtle energy but intuitively know where the problem is. Their hands or eyes go right to the spot on the animal or sympathetically to a spot on their own bodies. A lot of valuable subtle energy information is available if you will just watch their unconscious movements. While professing, "I just don't feel anything, Doc," they may be inadvertently scratching the acupuncture point GB 20.

Another excellent source of subtle energy information can be found in the client's small talk or their "story" of the problem. Intuitive insights that are not consciously perceived or understood by the client are often the stimulus for much of their spontaneous and volunteered conversation. Their mind is desperately attempting to communicate with you symbolically through their brain's chatter.

Information gleaned directly from the subconscious mind of the client can be a very accurate and helpful assessment because it has not been filtered through and censored by the rational binomial mode.

Clairvoyant sensitives are visually oriented and report graphic pictures coming to mind. These pictures contain symbolic information that must be interpreted by the sensitive. Symbolic information comes from the personal experiences of the sensitive and may not mean the same thing to anyone else.

Clairsentient sensitives will sympathetically feel the symptoms of the patient. This is actually a form of <u>brain</u> or thought reading and will include some of a human patient's binomial processing as well as the strictly analog data. Notice I did not say mind reading. These sensors will report on the "badness" of the condition, how long it has been going on and the emotions that are involved. All of this type of formation is binomial, based on categorization and judgments. This kind of sensitive will often empathize strongly with the patients and take on their emotional state. Many animal communicators I have met are accomplished clairsentients of this type. Unfortunately, they often neglect to strictly restrict their focus to the animal patient. When they come up with data that belongs in the left column of our dichotomy list, it must be coming from a binomial mind like their own or the owner's, not the animals. This is what we call anthropomorphism or mistakenly assigning the binomial thought processes to animals. According to Temple Grandin, animals are primarily analog, picture association thinkers. *V-2

The clairsentient sensor that is the most common and the most useful for my purposes is the sensor that sympathetically feels only the energy blocks. They do not pick up on the mental processing or impose meaning upon it. The actual energy block is often at some distance from the symptom being experienced by the patient. (The block causing a headache will frequently be found in the foot, for example.) There is a resonance established between the two bodies connected by focus, and the "sensitives" becomes aware of the places that do not match up or harmonize with their own. This is the most common form of analog perception, and the vast majority of people have simply learned, over time, to ignore it because it just doesn't seem to be relevant. These types of sensations are often dismissed with an almost subconscious scratch or rub.

I have found that almost everyone is sensitive to this information and can access it by simply following a few suggestions.

1. Sit down, relax, and breathe deeply into the acupuncture point

CV 6 just below your belly button. (centering point)

2. Remove your glasses or contact lens. Stare deeply into the patient. Do not focus but let the eyes drift apart until double vision is obtained.

3. Do not move. Resist scratching the itch, etc.

4. Do not pursue the information, let it come to you. Imagine yourself backing away if it doesn't come. Be totally passive except for focus.

5. Avoid language and naming parts of the anatomy. Simply point to where you feel heat, pressure, tingling, or involuntary muscle twitches.

6. Do not think about how much time it takes.

7. Do not judge the validity or importance of the information or try to understand it. Respond without thinking about it.

8. Do not take the whole thing seriously or worry about the consequences of being wrong. I will, as the doctor, ultimately evaluate the data according to the patient's responses so you can actually do no harm.

9. If it does not seem to happen for you, just pretend it does.

Keep a light, playful attitude. After all, what harm can it do?

To help the "sensitives" in this attempt, I will often physically simulate their CV 6 point. This brings the energy foreword and down into analog, Yin, or female.

When the "sensitives" indicate the location of the blocked energy, I explore distant acupuncture points with my finger or rumen magnet. I use classic Chinese theory and reasoning to guide this effort. For example, if they report a problem in the temple region, I will look at distal gall bladder points on the ipsi-lateral back leg, liver channel points on the contra-lateral back leg, triple heater points on the contra-lateral front, and heart points on the ipsi-lateral front leg.

The properly connected "sensitive" and the patient will both experience a release the instant an appropriate point is touched or focused on. When an abundance of energy is released suddenly, the sensitive may collapse or faint. That is why I want them sitting down. In less dramatic releases, the "sensitives" will yawn and report that the sensation is now moving along classic meridian pathways. Without any previous knowledge, they often describe the course of the meridian perfectly, right down to the ting (last) point.

I have used this technique with groups of sensors in many different demonstrations. Because it is additive, the more truly sympathetic individuals involved, the more powerful the effect. Research has shown that group prayer is also more effective for the same reason. *XII-6, XVI-15

Research at Princeton utilizing random event generators proved that both humans and animals can influence physical events with their intentions. Scientists showed that men are better than women at getting REG machines to do what they wanted, even though the women produced stronger effects. They also proved that male-female teams were substantially more effective at influencing the machines than individuals or teams of the same sex. If these principles hold true in medical treatments, the best results would be expected when a woman does the analog sensing and a male does the binomial diagnosis and treatment. In my experience this is the case. *IX-7

PROBLEM: If you will remember, the binomial mode has priority and can commandeer the energy away from the healing, analog mode. The analog mode is also dependant on the evaluations made by its binomial mode, in fact any binomial mode that is focused on it (Hypnosis). An individual who maintains strong binomial attachments and negative intent can stymie the healing process in themselves and others. This can happen, for example, when an owner is experiencing strong disappointment and frustration with the animal for its failure to perform. They maintain a judgmental, demanding, pushy attitude that is totally at odds with the healing process and will block it. In those cases, I will divert the owner's attention by sending them off on a mission, or I will come back later when they are not around or focused on the patient. I could also treat the patient later during my meditation session. Some individuals will perceive a religious conflict with what is going on and intentionally block it.

EXAMPLE: Water witchers and other diviners have been known to perform almost flawlessly in the field while alone or in the company of supporting believers. In tests set up and observed by antagonistic researchers they are unable to get consist results. This is considered, by the antagonists, to be proof that divining is bogus.

PROBLEM: "Sensitives" need to focus without judgment (witness) on the blocked energy and experience its subsequent release along with the patient. Some inadvertently assume a block or pain is "bad." This judgment forces them left into binomial mode and instantly closes down further present tense sensory input. The sensor is forced back into past tense or memory where things don't change; bad stays bad. Because they are now focused on the badness of the past, they do not sense and experience the patient's release in present tense. In binomial mode the focus is on separation, ego, or self. The sensor therefore often mistakenly identifies the analog message coming from the patient as their own condition. The patient experiences the release and healing while the "sensitive" is left holding on to the memory of their problem. When healers talk about picking up bad vibes from patients, this is what they are referring to. Bad is a judgment not unlike labeling, and it indicates binomial thinking.

Jesus said, "Do not judge, or you too will be judged. For in the same way you judge others, you will be judged." Matthew 7.1 Could he have been more specific?

In those cases where the sensitive misses out on the release, you have to either treat them directly using the same points, or get them to refocus on the balanced patient, present tense.

To avoid picking up analog data in the first place, stimulate points high on the Governing vessel (back midline) like GV 14 or 20 to push the energy up and back into yang, binomial mode. I like to call it the "Insensitive bastard mode."

ANIMAL SENSITIVES:

Horses usually have a fly-swatting buddy that they are bonded with. These two are focused mutually, and they strongly resist being separated from each other. They are on the same wavelength, so to speak, and it is primarily tuned in to analog mode.

I use this relationship frequently in treatments. Because these animals are bonded in analog mode, they will experience simultaneous releases. You cannot treat one with out treating them both, so I prefer to have them presented together. Some horses are apprehensive about being handled by strangers, and this pushes them into the flight and fight response or the binomial mode. In binomial mode, they protect their ego and block analog communication with you. Their buddy, safe on the sidelines, is still connected to his partner's analog mode and is less inhibited about communicating with me. Therefore I watch the buddy while exploring potential treatment points on the patient.

I can just treat the buddy if the patient is totally uncooperative and this would then be considered a form of "Surrogate" treatment. I use this form whenever the patient cannot be restrained safely or is too small to work with. A kitten or bird, for example, can be held by a compassionate. With the surrogate focusing on the patient, I simply diagnose and treat them in lieu of the patient. This can also be used in distant healing sessions (if the surrogate has a connection with the patient and can maintain a focus on them.)

Children are excellent "sensitives" because they have not yet learned not to be. They will also respond well to an authority figure's binomial suggestions that they can do it. However, most children do not have long enough attention spans to be of practical use as either a sensitive or surrogate.

MEDICAL INTUITIVES

"Sensitives," commonly called "medical intuitives," have been working successfully in the field of medicine with other doctors long before I began to make use of them. My specific contribution to this particular field of endeavor is to make it logical. Many of the "sensitives" I have read about are intuitively diagnosing <u>physical diseases</u>. That disease is then physically confirmed and treated by a doctor in binomial mode. Although the doctor's confirmation of the problem is certainly impressive and seems to prove their insight, it is actually illogical. Diagnosing <u>disease</u> is strictly a binomial mode function, and it involves separation, judgments and labeling that are not possible in strictly analog mode. The analog information in this case is ultimately being interpreted according to binomial principles and standards. We cannot logically mix checker and chess moves in the same game. We cannot logically mix Newtonian and quantum principles in the same strategy. Paradox, confusion, and complications are the result.

In this scenario, when the clairvoyant "sees" a tumor, the doctor will then most likely operate and physically remove it. This would constitute a successful treatment according to western physical standards and "prove" the value of intuitive diagnosis. A subtle energy assessment of the post-op situation, however, will invariably reveal that the imbalance responsible for the tumor remains unchanged. The treatment would be a failure according to energy standards. The tumor most likely will grow back, show up in other places or materialize as other problems in the future.

In this case, analog information is first translated into a binomial physical diagnosis. It is treated with a binomial materialistic strategy or fixed. That fix is followed by a second analog assessment or prognosis. Each translation between analog and binomial modes is subject to errors of interpretation. A binomial diagnosis of "cancer," for example, usually triggers an emotional response in the clairvoyant that blocks continued, present tense, sensory input. Focusing on the disease also distracts us from concentrating on the patient's over-all energy profile. The vital energy is split into past tense memories and future tense worry about consequences. Very little psychic energy is left in present tense to fuel an effective healing response. In other words, keeping track of all the different factors involved when we mix logical formats gets really complicated.

If, on the other hand, the original intuitive information is simply used to successfully rebalance the patient's energy field, the physical diagnosis and solution may not be needed at all. A well-balanced patient will often show a complete and permanent regression without any physical intervention. All the complications and confusion created by the translations can be avoided. Logical is simple. "Believe" is how Christ put it.

SUMMARY

Shifting focus quickly and cleanly from one processing mode to the other can be a problem. We can avoid this difficulty by creating a communication bond with another individual. One focuses on the energy body of the patient in analog mode, while the other maintains a binomial focus logically analyzing the sensor's data and formulating a treatment strategy. An assistant, the owner or another animal can be used for this purpose. Very small or absent patients can be positively affected by treating a surrogate who is focused on them.

WORKING WITH ENERGY

Yesterday I did seventy-four fertility exams, vaccinated forty-seven cows, treated three cows for indigestion, three for lameness, and operated on a cow to correct a displaced abomasum. No one expected or asked me to do any kind of healing work. It was a busy day and the element of time significantly influenced my decisions. I focused on the lower dimensions and functioned in binomial mode all day long until my meditation session that evening. I am reasonably good at applying binomial mode physical solutions using the cause and effect logic of Newtonian physics. I have been doing it professionally for thirty years.

Most of us are familiar with this approach. It involves muscle power, memory (education), doing to, efficiency, restraint, tools, instruments, chemicals, drugs, and vaccines. All these things are associated with the left side of the brain, materialism, binomial linear logic, and the left column of my dichotomy list.

If, on the other hand, I had decided to manipulate reality with energy, I have several options available to me. There are two basic forms of energy in reality: The explicated or measurable energy of the lower four dimensions, and the implicated subtle energy of the upper seven dimensions.

EXPLICATED ENERGY: Just how do you work with explicated energy? We can analyze and manipulate explicated energy with Newton's physics and relativity theory. Newton's technical definition of work is "moving matter or things." He defined energy as the potential to do work. He saw both matter and energy as temporal, occupying empty space. Energy was measured and defined strictly in terms of how it affected material objects. Energy that moves an object is called a force. A force, by definition, is limited to the lower dimensions where matter is manifest. Newton's physics did not conceptualize or define the kind of implicated subtle energy described in Quantum mechanics.

The basic force or energy in Western reality that can be measured, conceptualized, and manipulated is called magnetic or polar. Electromagnetic radiation and electricity are elaborations or transformations of it. It is made up of two opposite attracting qualities, poles, or charges. A negative charge attracts a positive charge and repels another negative one. The poles define a field of influence between them. True to their materialist orientation, Newtonian scientists see the magnetic force as a quality of matter. Because matter is immutable, the magnetic energy associated with it is also. It can be neither created nor destroyed. It is manifest at the subatomic level (dimension) as charged particles. The electron, for example, defines the negative charge and the positron the positive one. At the atomic level each atom of matter has a characteristic charge, which is determined by the particles making it up. Atoms combine into molecules based upon that charge. Two positively charged hydrogen atoms are attracted to and held by a negatively charged oxygen atom to form water, for example. In the chaos of normal chemical reality the poles effectively cancel each other out. Gross magnets are simply physical objects with their individual chemical polarities lined up. In other words, the chaos has been ordered in that material. The similarity of this polarity concept to the Chinese duality theory of yin and yang is obvious. Gravity is another basic force in nature that Newton could identify and accurately characterize but was never able to explain.

In the first dimension magnetism is represented as a straight line with a plus and minus at the opposite ends. In two dimensions, it is represented or graphed as a classic sine wave. The two polarities are indicated on the vertical axis as positive and negative values and time or movement on the horizontal one. Newtonian physicists work with these explicated or measurable forces in terms of the wave form's length, frequency, coherence, and amplitude. If I add two identical sine waves in phase, I will produce a wave with twice the amplitude or strength. If I add them one-half out of phase, the plus and negative cancel each other, and I essentially produce a straight line or no wave. The superimposed mirror image of a wave form also adds up to a straight line. Different waveforms and wavelengths can be added together to get an infinite variety of waveforms and therefore effects. *XV-2

WAVEFORM: This quality of energy determines the effect that energy will eventually manifest in reality. Schrodiger's formula includes all the different possibilities.

WAVELENGTH: This determines what physical structures in reality will absorb the energy. Frequency is often used to refer indirectly and inversely to this quality. The speed of electromagnetic radiation is constant, so a high frequency transmission has shorter wavelengths. Waves per second = Hertz.

AMPLITUDE: This is the amount of plus verse minus displacement from the vectored transmission line. It determines the strength of effect in reality.

COHERENCE: This is a measure of sameness or consistency of the radiations waveform and length. A coherent transmission has a narrow range of effect.

Simple wavelengths that are one to three multiples of each other will resonate or support each other. These wavelengths will have similar qualities or effects in reality. There are a total of seventy octaves in the whole electromagnetic spectrum so there are seventy potential resonant wavelengths. Each octave of that spectrum defines a different quality or type of reality for us. We recognize different ranges of wavelengths as microwaves, radio waves, radar waves, infrared waves, etc. Every octave range therefore has wavelengths within it that potentially will resonate with or have similar effects as those in every other octave. A color in the visible light octave can be correlated to a note in one of the sound octaves, for example.

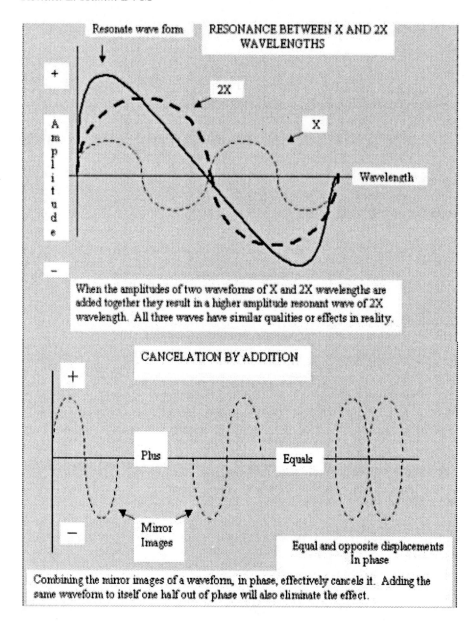

RESONANCE BETWEEN X AND 2X WAVELENGTHS

When the amplitudes of two waveforms of X and 2X wavelengths are added together they result in a higher amplitude resonant wave of 2X wavelength. All three waves have similar qualities or effects in reality.

CANCELATION BY ADDITION

Combining the mirror images of a waveform, in phase, effectively cancels it. Adding the same waveform to itself one half out of phase will also eliminate the effect.

Significant resonant effects occur only between double or tripled wavelengths. There is no measurable resonant effect between wavelengths that differ more than three times in length. To get any resonant effect between light and sound frequencies, for example, the energy has to first be transformed.

When we represent the full spectrum of possible wavelengths on a straight line, the 2x resonate points such as 5, 10, and 20 define progressively longer intervals or octaves. If I superimpose these resonating points by circling the line back upon itself, I get a clamshell like form of seventy nested circles on two dimensional paper or a spiral in three dimensions. Each circle represents an octave. The two-dimensional figure is amazingly similar to the one described by the dimensional theory in chapter two. The three dimensional spiral describes the movement of energy according to the theory of relativity. They are nested, which means that each dimensional circle contains all the dimensions below it. The outer circle represents the longest wavelength or the lowest pitched. The inner circle has the shortest or highest pitched wavelength. If outer circle is activated or vibrated, theoretically the next lower dimension or octave will be activated by way of resonance. This will in turn

activate a resonance in the next lower dimension or octave and so on

down to the lowest dimension. Energy moving from low to higher

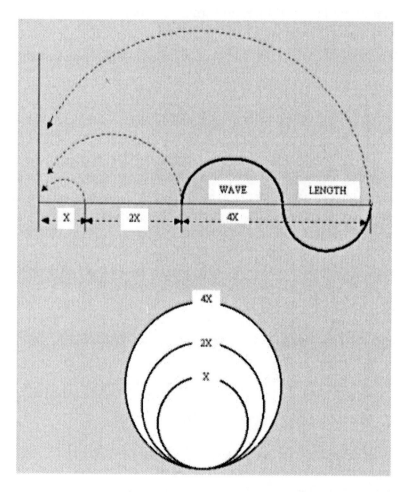

We get a diagram that looks like a clam shell if resonant wave lengths
are closed up into circles and superimposed.

frequencies (down) loses amplitude in proportion to the wavelength change. Therefore a waveform in the upper dimension will automatically project down into the lower dimensions and manifest as a higher frequency, lower amplitude event. Does this sound familiar? It graphically represents a mechanism that could transform a higher dimensional implicated idea or organizational plan into a lower dimensional explicated physical reality.

To continue this analogy, consider the fact that the outer ring or highest dimension of our model is the lowest pitched. The brain wave pattern that is associated with spirituality and the goal of meditation or prayer is the low-pitched delta wave. That wavelength is similar to the diameter of the earth. Our model suggests that the realm of spirituality is in fact outside or above the physical plain and therefore includes it.

We have three basic types of explicated energy that we can work with in medicine: sound, light, and magnetism-electricity. By definition, they manifest in space-time, and so we have to use Newtonian and Relativity logics relating to wave phenomena when we manipulate them. To maintain the logical format, we must constrain our focus to the four lower dimensions and the explicated physical material of the body. The logical conclusion or results of these manipulations will also necessarily be confined to the physical realm. We can "fix" the material body with these the same way we fix it with surgery, chiropractic adjustments, or drugs. There is no logical way, however, that we can directly affect the implicated process of healing with these explicated energies.

Sound (for humans) has about nine octaves of relatively low-pitched frequencies. They would correspond to some of the larger circles in our diagram. Light, on the other hand, is higher pitched and has only one octave. It would be represented by one of the smaller circles.

SOUND: Sound of course is the neurological interpretation of the electrical signals produced by a vibrating eardrum. The eardrum is moved by alternating layers of density (waves) in the gas molecules of the atmosphere. No atmosphere means no sound. These layers, usually produced mechanically by vibrating physical matter in the atmosphere, are transferred physically through the air as a relatively slow moving wave phenomenon. Sound waves can be used in medicine to move or vibrate physical structures in the same way they move the eardrum. We can select specific sound waves that resonant with the physical crystalline structure making up a stone, for example. If the amplitude is high enough, the vibrations created by the resonating energy waves can shatter the ordered structure of that stone. Protoplasm exists as a type of ordered crystalline fluid, and resonating sound waves can be used to vibrate this fluid into movement. Observations of living plant cells have demonstrated this fact. *XVII-1

Pathologies such as inflammation and pain have signature frequencies that can be altered with sound waves. The Chinese energy meridian has a specific length and a block in that meridian can be vibrated loose with a sound wave of the appropriate length. Acupuncture treatments can be viewed as "tune ups" or adjustments in the meridian's relative length.

Any physical structure in your body can be specifically vibrated by sound waves produced by your own voice. Vibrating a painful body part has an immediate anesthetic effect. You intuitively know this, and when a midnight trip to the restroom results in a stubbed toe, you will automatically try to search out the resonating tone that will disassociate the pain frequency by intoning "Ahhh," "Ohhh," "Eeeee," etc. You can easily establish this fact for yourself by simply touching or concentrating on a specific area while intoning different pitched vowel sounds or N and M. Research has documented that intoning the sound "Om" while in a meditative state significantly increases the blood levels of melatonin. That happens because those frequencies selectively vibrate the base of the skull where the pituitary gland is housed. You can use this idea to specifically stimulate any area or gland you chose in your own body or a patient's. To simulate an immune response, for example, you can vibrate the thymus in the area beneath the upper sternum. *XV-1 Because the sound that resonates with an area does not necessarily remain fixed, you have to experiment each time to find just the right pitch. A classic musical note covers a wide range of specific frequencies and it is not definitive enough to work predictably.

The moving eardrum and spiral cochlea of the inner ear transform the mechanical waveform of sound into electrical energy patterns for transfer to the brain. So sound can be used indirectly to induce or influence the mental and emotional states (mode). Individual sound waves, however, are two short to directly induce brainwave activity. But the brain can combine different frequencies coming into it from the two ears to produce wavelengths of the appropriate size. This is called brain wave "Entrainment." Combining a four hundred Hz wave with a four hundred and ten Hz wave results in a ten Hz entrainment wave; this would be an Alpha wave. *XII-4

Frequencies of sound are often represented for analysis on a circular format that can be said to represent one turn of the spiral cochlea in the ear. One of the seven octaves of sound is graphically displayed so that the resonating wavelengths of one to two are basically superimposed, as I did above in our model. The circle could start with a 256-meter wave and end with 512-meter wave for example. When perfect tonal qualities or intervals (notes) are represented around the circumference of the two dimensional circle, they end up overlapping, slightly, the starting point. This fact, called "Pythagoras' Comma," proves that energy does in fact move or exist as a spiral in four-dimensional explicated reality. The shape of the ear's cochlea is a graphic example. Sound cannot be accurately represented or analyzed with a two-dimensional circle. This same type of paradox occurs when our binomial brain attempts to superimpose a linear logical format upon the analog reality of the higher dimensions. *XIV-1 This circle is then used to graphically predict which frequencies can be combined for different effects such as resonance, additions, or interference. The logic is circular, in other words.

LIGHT: The electromagnetic wave spectrum has a total of seventy octaves. Light by definition is the only one we see. It is the forty-ninth octave with wavelengths from about four to seven thousand angstroms (small). Electromagnetic energy has wavelengths and is propagated at a much faster specific speed through space than sound is. That speed is the dividing line between the fourth explicated dimension and the fifth implicated dimension. Matter accelerated to this speed transforms into energy. Light exhibits qualities of both realities, manifesting as both a particle and a wave phenomenon. Physical scientists now believe this wave phenomena is part of the all inclusive "Quantum field". At death, our mind's focus moves upward from the explicated physical body of the lower dimensions to the higher implicated dimensions. We know this because on the way up our focus must pass through this transformation line, and people who have returned from the near dead consistently report seeing the light at the end of a spiral tunnel.

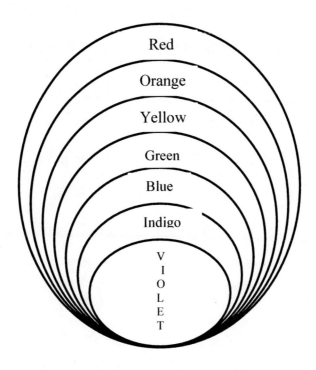

This is not a resonance diagram like some of the others. Colors all occur within the same octave of the electromagnetic spectrum; the forty-ninth.

Physical matter selectively absorbs electromagnetic waves or radiation according to its physical structure. An antenna resonates with or absorbs wavelengths according to its physical length, for example. The wavelengths that resonate with a particular antenna like structure in the body are absorbed, and the ones that do not resonate are reflected or simply pass through like X-rays. Atoms and molecules that absorb electromagnetic waves are "energized" into different states. Because this is explicated reality, quantitative qualities are relevant and important. We can warm a particular tissue or cook it, depending upon the dose of radiation. We use radiation in this way to selectively kill cancer cells, for example. The right amount of ultraviolet radiation absorbed by the skin stimulates melanin and vitamin D production, but too much ultraviolet burns the skin.

The logic and technology of manipulating electromagnetic energy is based upon the wavelength, frequency, coherence and amplitude. The wavelength of visible light is measured in microns so it physically resonates with the body at the atomic, molecular, and cellular levels. Radiated energy at these wavelengths is absorbed by the skin and does not penetrate far into solid tissues, so delivery to the vital organs can be a problem. Radiation levels sufficient to penetrate those deeper tissues will burn the skin. Light frequencies can be transferred to and stored in solutions or crystals either physically or with radionic devises. These frequencies can then be delivered to the internal physiology of the patient in the form of homeopathic or Bach like remedies. Light can obviously be delivered to the eye as a form of treatment. The full spectrum or octave of light is normally represented on a circle or color wheel. Complementing color frequencies occur opposite from each other on this circle. Color therapists use a circular logic similar to the sound logic. Colors, like musical notes and words, are categories abstracted from the continuous spectrum of reality and consequently include ranges of frequencies. Resonance is frequency specific and is difficult or impossible to obtain in many cases with these broad color classifications. *XV-3, 6

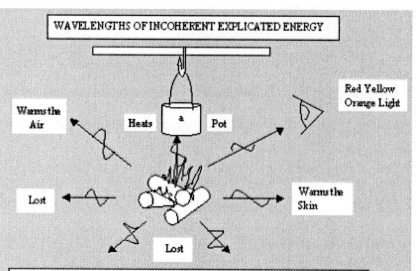

WAVELENGTHS OF INCOHERENT EXPLICATED ENERGY

Warms the Air

Heats a Pot

Red Yellow Orange Light

Lost

Warms the Skin

Lost

Incoherent energy radiated from a fire has many different wavelengths and therefore is absorbed by many different physical structures producing multiple effects. Radiated wavelengths that are shorter or longer are not absorbed, have no measurable effects, and are effectively lost. Coherent radiation = efficient

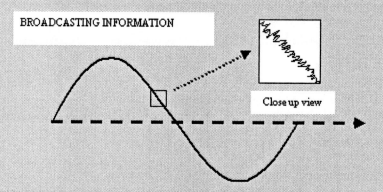

BROADCASTING INFORMATION

Close up view

A large broadcast wave of a radio transmission carries the smaller information waves. We can think of large explicated energy waves carrying the tiny implicated waves in the same way. Technically, however, implicated energy does not wave or even move.

Sound and light energy are both explicated and can be used to directly affect explicated physical structure. It is therefore logical to analyze and manipulate these physical effects with the Newtonian format. That logic is based upon cause and effect which dictates that we "do to" the patient or etiological agent. The orientation is to remove, eliminate, or cancel the perceived cause of disease, and we do this by selecting the appropriate frequency (note or color). The classic Newtonian linear format is modified in many cases, however, to include an element of circular logic.

MAGNETIC: Magnetic energy is implicated as far as our senses are concerned, and electrical energy is the explicated manifestation of it. There is no electricity without a physical conductor. Magnetism is yin and electricity is yang. Static magnetism or voltage becomes or transforms into moving electricity or amperes in "transformers." Magnetism occupies or projects through empty space but does not travel or have a speed. Magnetism is analog: It has no wavelength or frequency but it does have an amplitude or strength. We call that strength "voltage" or "potential." Voltage is the potential to transform into electricity or amperage. That strength of effect declines sharply with distance in empty space.

Magnetism itself does not move through space, but we can move magnetic poles and their associated fields. The effect of oscillating magnetic polarity changes in electromagnetic radiation travels at the speed of light through empty space. Moving a magnetic field in close association to a coiled metallic wire induces an electric current in the wire. In reverse, an electric current in a coil of wire will induce magnetism in an enclosed metal object.

Unlike electricity, magnetic energy fields are not constrained by physical reality. They pass through the physical body, and we can use them to affect changes deep within the tissues. We can use magnets in energy medicine to stimulate acupuncture points. *XV-4 Magnetic fields can also be used to stimulate bone fracture repair. *V-1 By manipulating the static polarity field in fixed therapeutic magnets, we can increase circulation and speed healing by activating the movement of ions within the blood stream. This is called the "Hall Effect." Strong magnetic fields can slow down the pumping of charged ions through the axons' membranes, and we can use that mechanism to effectively anesthetize a specific area.   *XV-8, 9

Pulsed electromagnetic fields can be applied using the same logic that

we use in electromagnetic radiation.   Pulsed magnetic field

frequencies are used, for example, to treat a number of pathologies

(such as inflammation and arthritis) by stirring up and destroying their

signature frequencies. The physical movement of magnetic fields can

be used to induce electrical flows in the physical structures of the

body such as nerves, blood vessels, bones etc.

ELECTRICAL: For most conventional medical practitioners,

"energy medicine" means using electrical energy.   Electrical

technology is used in medicine today both to treat and to monitor

patients. It is the basic form of energy used to power most medical

machines. It is perceived to be the energy of the central nervous

system and possibly the driving force of basic physiology. The spark

of life is considered by most conventional western Doctors to be an

electrical spark.   (Frankenstein brought his inanimate physical

creation to life with a jolt of electricity he captured from a lightning

bolt.) Generic electrical charges are applied in some cases to shock or

jolt the body with amperage.   Most therapeutic applications manipulate the frequency of oscillating polarity reversals to create a waveform. That frequency is conceived as the key and it is analyzed and logically applied the same way electro-magnetic radiation is.

Magnetism and electricity are energies that defy logical restraints. The electromagnetic phenomena straddles the dividing line between implicate and explicate realities. As with light, they do not fit entirely into either the Newtonian or Quantum formats. These energies are typically applied according to trial and error experience.

IMPLICATED or SUBTLE ENERGY: In the multidimensional reality described by scientists, all the separated and sensed (measurable) things of the lower dimensions are actually parts of a total energetic unity. In the highest implicated dimensions there are no separated identities. The lower implicated dimensions of matter and explicated energy however contain their individualized organizational blueprints. We call that pattern or information "their subtle energy profile." This energy of analog reality is all-pervasive, everywhere and nowhere at the same time. In these dimensions, there is no space or time, so there is no separation, location, sending or transfer. Energy at this level has no wavelength to work with. In the lower implicated dimensions subtle energy can only be said to vibrate. In the Celtic and Chinese traditions, this vibrating energy is divided into a handful of subsets, and the Chinese five-element theory describes how these vibrating subsets are related. True to their analog perspective, they place them on a circular format like the ones used with sound and light energies. The elements are then manipulated according to that same circular logic. For example, elements that occur across from each other on the circle have a powerful balancing or resonating relationship.

The western world's super string theory describes reality at this level as vibrating strings closed up into <u>circles</u> or, more precisely, as pulsating <u>spheres.</u> The wavelength phenomenon of space-time reality has been dropped, but elements of frequency and amplitude remain as vibrations. Imagine a jiggling mass of Jello. *XV-5 This united vibrating reality is like a symphony. God is the composer, and the dandelion is one of the instruments in the orchestra. The dandelion's vibration is an integral and important part of the total composition being played, but its individual isolated contribution is not normally perceived. If everything is entangled as quantum mechanics indicates, we are potentially experiencing the vibrations of the whole symphony in which every individual instrument is a part. We are already connected to every analog vibration in our world, in other words. The implicated reality that defines our individual being is awareness and the ability to choose and focus. We call this the mind or consciousness, which should not to be equated with its lower dimensional physical manifestation called the brain. It is the mind that determines, by way of focus, which vibrations to experience, block, or ignore within the whole symphony of reality. *XV-7

Subtle energy fits exclusively into the Quantum mechanic or Super-string formats and should be applied according to those tenants. I use Q.M. because I don't understand S.S. well enough yet. Unfortunately many practitioners of energetic medicines confuse the issue by indiscriminately mixing in Newtonian principles.

Focus is organized around the common sensory inputs of lower reality that the brain can relate to. For example, when the mind's focus moves up at dying, the individual first "sees" the light and then "sees" its body from an elevated position. The truth, however, is that the mind or soul perceives this phenomenon at the implicated level of reality where there is neither location nor eyes to see. The mind simply communicates the idea down to the brain in a form that will make "SENSE" to it.

The healer's job is to some how get the patient's mind to focus on the particular subtle energy information that will balance his or her energy profile. We usually have to do this by way of their brain's sensory inputs and conditioned associations. The vibration we need is always ultimately implicated and out of sight, so we are forced to use the explicated manifestations their brain normally associates with it. We simply use the sight, taste, smell, feel or sound of an explicated energy or physical remedy as a way to ultimately focus them on the implicated vibrations of its analog subtle energy plan. We must keep in mind, that the sensory experience is only an abstract of this reality. Let us not confuse the map with the actual terrain, or the symbol with what it stands for.

ANALOGY: In the reality of broadcasting we are working strictly with energy. All radio and television stations are broadcasting different electromagnetic wave bands, at the same time, which we can not perceive directly. All the channel frequencies simultaneously

267

pervade the space around us as potential information sources (parallel universes.) The one that manifests for us is the one we tune in to with our receiver (sensory organs). In terms of Q.M., we collapse the total set of possible vectored wave functions (channels or stations) to one when we tune in to it. The finer more detailed information (implicate) that we are actually interested in is carried on those larger broadcast waves (explicated). The information (reality) we receive is therefore ultimately determined by our tuning (focus). However that information or reality is actually independent of that particular broadcast wave and can not be equated with it. I could get the same information from a newspaper. It is the information that ultimately is important and causes change, not the messenger or media of exchange. Believe it or not, it is possible to broadcast nonsense.

Consciousness is the highest and most inclusive form of implicated subtle, energy. The part of consciousness that makes decisions is called the mind. The conscious mind of the doctor/healer can also be used to heal either themselves or a patient. The healer's mind is connected via attention or focus directly to the implicated energy of the patient (subtle body → subtle body.) Complicating transformations and interpretations are unnecessary or minimized because the transaction is basically intra-dimensional. This communication is on a very high and inclusive dimension so it can and does result in powerful physical healings in the lower dimensions. On the mind's level of reality there is no cause and effect relationship to <u>work</u> with. Intent is the higher dimensional and more inclusive manifestation of that idea. Intent is dependent upon belief. The intent to walk on water must be predicated on the belief that it is possible. Belief is one of the basic spiritual lessons we must learn on the way to enlightenment. **Enlightenment** is focusing on and accepting the reality and logic of the higher dimensions. The consciousness is fortunately an understanding entity that allows for the individual healer's level of enlightenment. It will manipulate (censor and edit) the perceptions so that they ultimately make <u>sense</u> to its own lower dimensional intellect. The simulated physical details of a given situation therefore reflect the unique histories and enlightenments of each individual healer. Clairvoyants will describe different colors and clairsentients will feel different energetic distortions in the same patient, for example. Your consciousness will supply you with something to physically fix if you need that concept to sustain a focus and belief. The perceived details of this subtle energy reality are individualized for that particular observer and may differ substantially from others' perceptions. Each interpretation is simply one of the wave forms in Schrodiger's equation. This is an expression of the Quantum principles of sum over paths (parallel universes) and probability. The different interpretations of that reality are all true or real for that individual. Ultimately the loving intention to heal, belief, and focus are all that is necessary to successfully work with subtle energy.

Many gifted healers like Rolling Thunder do not work at it professionally. If you make your living healing, and depend upon the fees to pay the rent, it is difficult, if not impossible, to maintain a purely altruistic intent and focus. This is obviously the problem that most plagues the business of modern medicine today.

ENERGY MEDICINE

We can manipulate explicated energy forms using a classic Newtonian format to affect changes in the physical structure and chemistry of the body. Radiation can selectively destroy pathogens or unwanted tissues such as tumors. Sound waves can pulverize stones or disassociate pain. Magnetism can move ions within the tissues and increase circulation and bone repair. These external energy applications "do to or fix" the perceived physical problem. They are often used logically to negate, remove, or subtract a perceived cause, symptom, or etiological agent. Because large quantities of energy are usually applied when the Newtonian logical format is followed, binomial parameters such as quantitative dose are of critical concern in these technologies to avoid inadvertent damage or poisoning. Healing can be supported in this way, but not directly influenced.

We can, on the other hand, use the implicated aspects of those same explicated energies or of physical matter (remedies) to directly affect a healing response. But we must apply them using Quantum mechanic logic and principles. In quantum reality every thing is connected, and that includes the explicated and implicated qualities of energy and matter. We cannot separate the sound wave from the subtle energy of the sounder, the light from the subtle energy of the radiating material, or the information from the intent of the sender.

The implicated aspects of that energy or physical materials will affect the patient's energy profile on the higher, more inclusive, healing levels. If we are going to use the implicated aspects of one of these explicated energy forms or physical remedies to manipulate a healing response, we must carefully select the energy source. The focus is always on the subtle energy component of the physical, vibrating, or combusted material, not the chemistry, sound, or the light itself. The light from an incandescent bulb carries the energy of tungsten. The sound of a cedar flute carries the energy signature of that plant. The heat waves of burning moxa carry the energy of mugwort. See the chapter on sound therapy for more details.

Healing, like learning, is a function of the implicated patient and cannot be physically forced or imposed upon them from an external explicated energy or material source. The natural flow of effect is top → down. To attempt to influence the higher implicated healing response with lower explicated energy qualities such as generic colors of light, frequencies of sound (notes), or physical remedies is not logical. Individuals using these explicated generic energies in healing work can and often do document some impressive successes. The healing response is due, however, to the focused intent and beliefs of the healer and patient and not directly to the explicated energy. In those cases, the color, note or physical remedy is used as an abstract symbol for a particular energetic phenomena such as an Indian chakra, Chinese element, or meridian. They are simply symbols that direct the attention of the mind. They are, in fact, a placebo.

A placebo is a very affective and useful tool in conventional materialistic medicine if it is chemically and metabolically innocuous like sugar. In energetic medicine a placebo should be as energetically neutral as possible. If it is not, the implicated energetic profile of the placebo material can ultimately interfere with or even over-ride the original intent of the treatment. A patient in need of the earth element (Chinese medicine) or third chakra simulation (Indian medicine) is usually said to need the color yellow. However, that patient may not need the energy of cadmium, which is typically used in making yellow filters and pigments. A patient being purposely stimulated with a particular musical note from a wooden flute may actually be allergic to the plant that the flute is made of. *VI 5

The implicated aspects of explicated energies and remedies can be used to directly influence, by addition, the organizational plan or template of the patient. It is those plans, after all, that ultimately control and produce the healing process. Energy patterns used to influence such an organizational change are called information, the transfer of information is called communication. A true "healer" is not a technician doing to a patient but a communicator, guide, or teacher offering information and suggestions. The patient or student, however, is ultimately responsible for tuning in to that message and accepting or rejecting it. The higher dimensions of reality are the most inclusive, and therefore communication that is accepted and included at these implicated levels result in more profound and lasting healing on the lower physical plain.

Hands-on healers use their own subtle energy, or Chi, to manipulate the energy body of the patient. The intent is to negotiate a balanced flowing state by removing perceived blocks and/or imbalances. Qi Gong, Reiki, Therapeutic touch, and other healing modalities like Barbara Brennan's all use physical hand movements to direct their corrective measures. Lawrence LaShan classified them as "Type 2" healers. Practitioners using these different methods are often most effective healers. They each have entirely different and complicated technical explanations for what they are <u>doing</u>, however. That is because they are all attempting to apply cause and affect logic to a dimension where it does not actually exist. Complications always result when perturbant adjustments are added to a simple basic logic. They are attempting to understand something in terms of physically sensed vision or touch that technically does not actually manifest there. The technical details of these modalities are interpretive fabrications of their own individualized consciousness.

Spiritual healers use the energy of their minds to affect a healing response while maintaining a meditative or prayerful state. Physical movements are usually suspended and the vital energy is invested entirely in the mental dimension as focused attention and intent. Robert Stone calls the healing technique of creative visualizations used in these practices "Cyber-physiology." Prayer, the Silva method, and Lawrence LeShan's "Type 1" healers are examples of this type of healing energy manipulators. The fact established by Stone and others like Bernie Siegel who have studied this phenomena, is that the visualizations do not have to be realistic representations of the perceived physiologic mechanisms involved. In fact, too much detail in the visualizations often distracts the focus from the goal of healing. Simple is usually better. In fact they can be totally fabricated and fantasized. The visualizations obviously are effective only because they define and clarify the all-important "intent" of the mind. * III 1, 6 *XVI 4, 11

EXAMPLE: Suppose an air traffic controller realizes that a plane is on a collision course with a mountain. Interventions could be designed, using Newtonian logic, to <u>physically</u> deflect the plane or remove the mountain with something like a guided missile. Using this same logic, we could also try to <u>energetically</u> deflect the plane or destroy the mountain with lasers, ultra-sound, or a magnetic device. This technology is theoretically possible but has yet to be perfected. The relatively large amounts of external energy inputs needed make all these approaches dangerous. If that energy is not carefully and precisely applied, these solutions can cause serious negative consequences.

A third possibility would be to use energy according to Quantum Mechanic logic and principles. In this logic change is only achieved by way of additions. We need to alter the implicated flight plan by adding information to it. We can communicate with the pilot via electromagnetic waves. If the plane has a radio receiver and it is tuned in to the right frequency band (focused), we can inform the pilot of the problem and suggest a solution. The solution is an alteration of his original implicated flight plan. The pilot could then turn the plane accordingly, using his own intrinsic mental and muscle energy. The plane's own fuel and engines would power that change of course. This is ultimately a much safer approach because the small amount of external electromagnetic energy involved in the information broadcast would not likely damage anything physically if it were improperly or imprecisely applied. It is the precise quality of the energetic message that is critical in this case and not its quantity or volume. The success of this approach, however, is dependent on some intrinsic factors of the pilot and plane. If he cannot or will not tune in to (focus on) the communication, the information will not be perceived and acted upon. The pilot cannot act upon the information if he does not understand English. To execute the corrective measures he must have fuel on board and the plane's controls must be functional. We cannot control or influence any of these intrinsic factors from our position the ground. This communication approach will work fine if the pilot's original implicated flight plan is to arrive safely at their destination. However if he is a terrorist bent on destroying the plane and its passengers, the information will be ignored. We cannot possibly change that higher implicated plan or program directly with physical matter or explicated energy. Subtle implicated energy patterns (ideas) are the only way to affect a change on that level of reality (spirituality).

SUMMARY

There are some basic differences in how we work with energy in medicine as opposed to physical things. Explicated energy medicine usually involves some basic manipulation of the energies fundamental wave properties of the frequency, wavelength, coherence, and amplitude. Sound and light are two explicated energies we can use in these ways. They can be applied in medicine using the same basic linear Newtonian cause and effect logic we use with physical manipulations. However, in many cases, practitioners use types of circular logic such as the color or octave wheels to predict and calculate additions, resonance, and interference.

Magnetism and electricity do not fit entirely into either the Newtonian or Quantum formats. They are consequently used primarily on a trial and error basis.

Explicated energy has an implicated element that we can also work with. The subtle energy signature of the material producing the sound or light is carried by that explicated energy the same way large radio waves carry the more subtle information component of a broadcast. This information is contained within the intrinsic vibrations of the larger wave form. This implicated information must be applied according to Quantum Mechanic principles.

Implicated or immeasurable energy can only be said to vibrate. Because there is no matter to move in the higher dimensions of reality, the Newtonian concept of force and cause and affect are not manifested. These vibrating patterns we call information and its application we call communication. We might also call them prayers, songs, logics, organizational plans or template. Information communicated to and accepted by the patient's higher dimensional reality can potentially reprogram defective organizational plans. Defective organizational plans cause disease by interfering with or distorting the accurate projection of that patient's own implicated purpose into physical reality. Profound healing occurs when the organizational plan is properly adjusted to allow the basic implicated purpose (Soul) to manifest properly.

The implicated energy we call the mind can be used to heal as well. The mind manifests with the master plans in the higher implicated dimensions and can therefore communicate with them directly (Subtle energy → Subtle energy). At this level of reality (spiritual) we have only focus, intent, and belief to work with. Simple is good.

ANALOG MEDICINE

My goal is to promote a logical approach to the practice of medicine. Conventional western medicine has and is still effectively applying the binomial, linear logic of Newtonian physics to the explicated, physical dimensions of life. I have, consequently, called this "Binomial medicine." Unfortunately, they have also tried to impose this limited materialist strategy upon the higher, implicated dimensions of the life process, where it does not apply logically. Healing, for example, is a higher dimensional energetic phenomenon.

Analog medicine" is what I call my attempt to apply the logical principles of Quantum Mechanics in the practice of higher dimensional medicine. A strict adherence to a logical format promises to transform the present confusing situation in what is commonly called "alternative medicine" into a comprehensive science. An energetic medical science will more easily integrate with, and compliment, western medicine's materialistic, scientific approach. I chose the logic of quantum mechanics because it describes a strictly energetic reality, it is the most productive logical format ever developed, and I think that I understand it well enough to work with it. Super-string theory is, however, another logical possibility. I chose to call it "Analog medicine" because the analog quality embodies the most fundamental and characteristic aspect of reality at the quantum level. It is paraphrase of the basic quantum principle of entanglement or connectedness.

The idea is a simple one. Design a system of medicine that conforms to quantum mechanical principles in each and every step of its protocol.

DESIGN: Entanglement means that the doctor cannot be separated from the treatment or the treated. The practitioner's personal energy balance, experience, abilities, constitutional type, and intent must all be taken into consideration in the design. Individuals differ in the way that subtle analog energy manifests for them. We each have sensory prejudices that dictate whether we see, feel, hear or just intuit this energy. We are all at different levels of enlightenment.

This is obviously contrary to the conventional medical approach, where the procedure and medicine are standardized. In analog medicine, what works for me won't necessarily work for you. The principle of sum over paths says that there is no one correct way. We can logically use different specific methods as long as they are analog by nature and work.

EXECUTION: Classic medical protocol typically follows these steps:

1. Examination   2. Diagnosis 3. Prescription   4. Treatment

5. Prognosis

Analog medicine (as the name implies) is connected or entangled. It does not have definitive steps like these. The act of finding and focusing on an effective balancing point or remedy, treats the imbalance at the same time. The release that I described previously as the ultimate goal of treatment is, at once, the diagnosis, prescription, and treatment. There is no future tense or prognosis in quantum reality.

Expository language is of course strictly binomial, and so we are faced here with the classic translation problems of describing in binomial terms a strictly analog reality. This is the same problem you will have trying to verbally describe a work of art. It really is impossible, and sometimes even difficult.

Most of the alternative medical modalities that I have studied contained some good and very useful analog elements in their protocols. My strategy has been to identify those elements and then combine them into one totally analog system of medicine. (See Important Principles I Learned)

EXAMINATION: In analog medicine, we must always keep in mind that we are evaluating the energy body or the implicated reality that ultimately manifests the physical body of the patient. The uncertainty principle of quantum mechanics maintains that the act of measuring energy changes it. Engaging a patient with our focused intent creates a healer-patient entity. It is actually this combined entity that is evaluated in the end. Another doctor's evaluation of the same patient can differ from ours and still be legitimate. I could say that the right side of a teeter-totter is too high, while another doctor might conclude the left end is too low.

Symptoms inevitably guide and influence an examination. They usually are, after all, the complaint and reason for our interaction with the patient. Symptoms are, however, judgmental categorizations and therefore strongly binomial. If we focus on them, they will automatically shift us left, out of the healing mode. Symptoms are like binomial road signs. We really need them sometimes to guide our trip, but if we fix our focus on them too long the trip may end up in the barrow pit.

Analog examinations are all basically different forms of divination or dowsing because this reality is implicated. That is how we access information from the illiterate analog mode. (See Divining)

Some of the proven divining techniques in alternative medicine include: Acupuncture pulsing, Ah Shi point palpation, Shu point palpation, radionic stick plates, muscle testing, Qi Gong, Reiki, Voll instruments, Therapeutic touch, sympathetic sensitives, Ting point analysis and interviewing.

The physical senses of olfaction and touch are the most analog. Odors do influence me but I am not usually consciously aware of them. I do a lot of palpating of Ah Shi, Shu and Ting points. When remedies are indicated, I use the stick plate (touch) of a radionic machine. I keep in touch as much as possible.

In analog medicine, the principle of entanglement dictates that the doctor and patient's energies be connected. Communication is another word for this entanglement or connection. Maintaining two-way communication is an essential element at every stage of the treatment. Consequently, I prefer examination techniques that rely upon constant feedback from the patient. I encourage this by asking questions both verbally (humans) and physically: "Is this point sensitive?" The animal patient responds to these questions in many subtle ways. (see "Release" in Health and Healing) This approach intimately engages the animal and human patients in the process, both mentally and physically, and they become active participates, rather than just passively submitting to the process. Bernie Segal found patient participation to be a critical factor in their survival. *XVI 11

In my experience, energy appears to move along the classic Chinese meridians. I tend to see the Yin/Yang and five element relationships unfolding in energetic reality. The logic that I use primarily revolves around balancing meridian energy according to the ancient Chinese classifications of time, name, neighbor and element. Doctor Richard Tan teaches this approach, and he calls it four point balancing. *XVI-25, 26

Other people will, just as effectively, see and think in terms of the

East Indian Chakra system or the aura.

DIAGNOSIS: This is undoubtedly where analog medicine departs most radically from conventional medicine. The diagnosis of disease is the bedrock and central element of western medical logic. Western medicine treats diseases, and no diagnosis means there is no meaningful treatment, only symptom suppression. The primary emphasis is on removing, eliminating, or neutralizing the perceived cause of that particular diagnosed disease.

Diagnosis in analog medicine is a statement describing the state of health or balance in the individual patient's energy field. There is no space or time in quantum reality, there is only present tense. Therefore there can be no cause or etiology. Energy medicine can deal only with the individual patient's energy body in present tense. It is either balanced, or it is not.

This is one reason why I like the Chinese system. I can describe a diagnosis in terms of an individualized statement of balance without any labels, names, or categorization. Treating the individualized patient and rebalancing their energy profile is the same thing. This keeps my feet solidly in analog mode.

PRESCRIPTION and TREATMENT: Rebalancing the energy field is what I do in Analog medicine; in fact, it is the only thing I can do.

I learned several excellent analog treatment protocols in the alternative modalities that can be used to effectively rebalance a patient's subtle energy. They include: Bach flower remedies, Reiki adjustments, Qi Gong adjustments, magnetic therapy, Therapeutic touch, Zero balancing, Radionic broadcasts, Homeopathics, sound therapy, meditation, Prayer and moxibustion.

Some reasonably good analog treatments are included in the disiplines of Acupuncture and Herbology.

I personally use many of the above treatments. I decide on what method to use while diagnosing the particular case. Many things influence my decision. I would not use AP needles on a wet, dirty animal, for example. The available time, resources, history, and facilities can all influence this decision as well.

Ritual is a significant element in medicine and it often determines what I use in a particular case. For example, some horses get upset around needles, the smell of alcohol, or when restraint is applied.

Because I usually engage the owner/handler in the treatment, their own past experiences and beliefs must be considered. The owner's anticipation of results is extremely important, and so if they like and believe in bells and whistles, I try to oblige. Some of the better treatments (like QI Gong and Zero balancing) are primarily mental and involve very little or no physical effort or instrumentation. Western clients traditionally focus on and consequently value material and force. They are not comfortable paying someone for <u>not doing</u>. I try to put on a good show for those people. I have a laser and infrared stimulators, a Chi machine, and pulsed magnetic field stimulators. In other words, I provide them with a ritual they can believe in. I have found that some owners have a curious fascination for needling and just simply want to see that done.

When it comes down to me and the patient one on one, I use a Qi Gong/ Zero balancing kind of acupressure until I get a release, and then follow up with a homeopathic remedy of a local plant. I intuitively select a plant or use a radionic machine to divine one from my personal tinctured collection. (I do not use sound therapy very often on animal patients.)

I use local plants as remedies because they have a more intimate and stronger connection to both the patients and me. I also want to collect the plants personally after first experiencing them growing, undisturbed, in the wild or in the garden. A period of focused, meditative contemplation down loads their energetic signature into my analog subconscious for future reference. This occurs even thou I am not consciously aware of its energy profile at the time. Therefore, I have a personal analog memory of every plant in my collection. I can use this memory of the plant in lieu of an actual specimen. A remedy can be applied by visualizing the living plant and its surroundings, while staring at the patient. I use whole plants or representative parts of the entire plant in my specimen bottles whenever possible because concentrating on a part would be binomial.

In analog medicine, the divining procedures used during the examination and diagnosis phases obviously access only the energetic qualities of the plant remedies. They do not evaluate their physical chemical qualities. Many very effective energetic plant remedies are chemically poisonous. It is only logical to use them as homeopathic preparations. I use a "Tansley Direct Transfer Stimulator" or a radionic instrument like the "SE-5 Biofield Spectrum Analyzer" to prepare these.

A homeopathic plant remedy does not add to the patient's system in a quantitative way as material medicines or remedies do. Western medicines and herbal remedies actually supply physical materials or chemicals the body may need. The quantified dosage is therefore of critical concern to ensure effectiveness and avoid poisonings. The plant remedies I use, on the other hand, do not contain the chemicals of life. They are qualitative. They ideally contain only the plant's implicated plan, pattern, frequencies or song, if you will. The full message need only be delivered intact to be effective, its volume is irrelevant. Resonance in the patient's system will amplify the message appropriately if it fits.

A plant's energetic profile contains the specific strategy or logic it uses to survive in the world. That plan is expressed in the plant's own particular balance of the five basic elements. A desert plant, for example, has developed ways to compensate for the excessive heat and lack of water. It is this specific implicated pattern or strategy, if you will, that we are offering to the patients as a solution for their particular problem. I am in effect saying, "Try this way of dealing with life instead. It works for the plant."

In analog reality, plants are like people and animals. Within the genus species, there is a general similarity. However individuals, even those from the same planting, can vary considerably from the average. The specific growing conditions of the plant can affect its energy profile. The soil type, competitive neighbors, age, chemical exposures, and weather are just some of the many factors that can alter the individual plant. Consequently, you cannot assume that every Aspen tree is exactly the same energetically. Rolling Thunder routinely identifies very specific individual plants from within an apparently homogenous growing population for his patients. *I-2

PROGNOSIS: In quantum reality, there is no future. Balancing a patient's energy field is a present tense phenomenon, and it cannot cause a future tense event. Balancing puts the patients back in touch with their implicated master plan or soul. Once they are balanced, all you can do is get out of the way and wait to see what manifests. Master plans contain elements that are often considered to be disease.

Pain, aging, and death are in the master plan of all multi-cellular organisms, and resisting them, when the time has come, only creates unnecessary tension in the system. It is the resistance itself that actually creates the disease. Analog medicine can only guide the patient to re-manifest their master plan. It cannot change or alter that plan.

ENGAGING and MANIFESTING: In the implicated reality of analog medicine, you are manipulating only the patients' energy field. You are doing this to orchestrate a healing and a subsequent physical change in their body. You must keep in <u>mind,</u> however, that there is no cause and effect relationship at this level. Your relationship is more as a teacher guiding their efforts than doctor doing something to the patient. It is ultimately by way of the student/patient's own energy and effort that the learning/healing manifests, not the teacher's/doctor's. Balancing an energy field does not cause healing, it only allows for it, a subtle but important difference that is often overlooked. The doctor must engage the patient in analog mode, negotiate a release to achieve a balanced system, entirely disengage and wait in judgmental, binomial mode to see what manifests. It is a mistake to stay energetically engaged after a successful release is obtained.

Energy suddenly released from a block will cause turbulence as it flows back into the general circulation. This is often perceived as negative sensations in other parts of the body. Pains will move, for example. It is counter productive and a waste of time to chase down these new symptoms and try to eliminate them. Additional energy adjustments made at that time will only confuse the original one and disrupt the balance that was originally achieved. The new balance needs time to physically manifest, and the doctor should not intervene again until progress seems to plateau.

In some cases physical changes can occur rapidly or even instantaneously, as they did for Jesus. In most cases, however, you will see rapid symptomatic improvements but the physical changes will take considerably more time. Aggravation of the symptoms can also occur before positive changes are apparent. This is called a "healing crisis." New symptoms like skin rashes, diarrhea, or flu like aches and pains may be precipitated when a major "detoxification" occurs. The healing crisis and detoxification are both unpleasant yet positive responses that proceed true healing. Healing is not always fun.

You are planting a seed of healing and it needs time to sprout and grow. Don't plow it up with a second planting before it has a chance to manifest. I see that frequently in alternative medicine because treatments are often applied without monitoring their effects. Our binomial left side is accustomed to equating results with effort, and so in that brain's logic, the more we do the better. The binomial brain also perceives a future and becomes impatient waiting for results. A patient who has been successfully balanced with Reiki will often then be given a homeopathic or some other powerful energy treatment before the Reiki balance can manifest. This confuses the issue in the doctor's mind and in the patient's energy field. The result may be that the two perfectly good energy treatments cancel each other out.

ANALOG TREATMENTS

EXAMPLE: When I acupuncture a horse, I try to first focus on the whole animal, ignoring the client's complaint as much as possible. Analog is a passive state, so I relax and center myself while taking notice of intuitive hunches and clairsentient feelings in my own body. On a good day I may actually sympathetically feel stinging in the treatment points. I stay tuned into this mode throughout the procedure. Conscious attention is the unifying connection I need to work with. To first establish a communication dialog, I will direct the horse to do a couple of basic physical maneuvers such as backing up and stepping to the side. I now have the horse's attention and proceed to ask it diagnostic questions. Touch is one of the most analog of our senses and I stay in touch throughout the whole process. Diagnostic points are palpated. I survey the association points on the back for pain and the ting points on the feet for puffiness. I watch and feel the horse's responses to each of these points. That survey establishes which meridians are involved in the basic imbalance. With the owner's complaint in mind, I examine the pathways of the meridians indicated. When an appropriate point is engaged, the horse will respond with an obvious "release."

In analog mode I am simply communicating mindful intent to help. I am listening to his opinion, asking permission to treat, and agreeing not to impose. Physical and mental connection is maintained at all times. I only needle points that result in obvious releases upon palpation. The horse rarely moves when these specific points are needled. Very few points are needed in analog treatments, and no restraint is required usually. Some painful needling sensations will actually be tolerated by the horse as long as we maintain the mindful communication link. The horse ultimately associates me with the good feeling of his release and will seek me out in the future.

After one or two successful analog treatments, the horse becomes conditioned to the routine and will automatically release when I first contact him. "Hurry up and do something Doc, he's starting without you."

By contrast, in binomial mode I <u>do</u> <u>to</u> the patient what I think will be good for it according to my <u>past</u> <u>tense</u> experience with other patients. My solution will be <u>forced</u> upon him, whether he agrees or not. I will likely try to impose a <u>standardized</u> point prescription for his <u>symptom</u> picture or disease. A horse can easily sense the energy sent into my muscles before I actually execute such an aggressive binomial movement. He will automatically respond by shifting left into classic binomial flight and fight. As you will remember, that left shift automatically shuts down any further analog input and disassociates our communication link. The unity (connection) of analog mode is lost, and I will have to resort to some serious physical restrains. The horse dislikes the whole experience and will avoid me at all costs in the future.

EXAMPLE: I can balance patients with analog plant remedies and usually follow acupuncture treatments with them. A blood sample or Polaroid picture is used as a witness in the radionic analysis. The remedy is selected in a blind survey of my specimen stock solutions. Memory and past tense are binomial, and therefore my previous clinical experiences with the plant are not relevant to the present tense patient. Neither are the herbs' established physical medicinal qualities.

The animal's owner, a clairsentient, or human patients themselves can do this survey physically by passing a hand slowly over the specimen racks. Initially I preface their attempt with the suggestion that everyone can do it. I proceed with a carefree, playful, and matter of fact attitude to dispel any performance anxiety. Heat or tingling sensations typically indicate resonance. A human patient or a sentient, sympathetically experiencing the patient's energy block, will realize a release when their hand passes over the right remedy. The intensity of pain is also a very sensitive and accurate criterion to use in these surveys. We usually find two or three plants that strengthen the patient's general vitality frequency as well as eliminate or cancel the frequency of one of the primary symptoms.

I preferentially use the <u>Tansely</u> <u>Direct</u> <u>Transfer</u> <u>Stimulator</u> to make my homeopathic remedies because it has an analog potency dial. Many of the other devices use digital ones. I dowse the potencies of transfer as well. Most remedies require several potencies.

This homeopathic remedy is given to the patients as needed. It can be taken orally or dropped onto the points used in the acupuncture treatment. The response is an immediate and obvious release when it is needed. Their positive response to the sight and touch of the remedy bottle determines if it is in fact needed. They will eventually experience the same kind of conditioned response I described above with acupuncture.

EXAMPLE: The sound of a remedy can be used to treat a patient in analog medicine. The sound of a balancing frequency quickly results in a release of tension and a profound sense of well being. Pain and other symptoms are immediately mediated or completely eliminated. Therefore a trial and error sound survey of the possible potential remedies available is a way these can be selected. I can actually play one flute at a time or have them listen to a series of recorded flute sounds. I can, instead, add the fresh or tinctured specimens of potential remedies to a resonating bowl. However, sound responses correlate well with the radionic evaluation of the physical specimens. Therefore, I can simply pick sound remedies from the list of potential homeopathic remedies previously dowsed with the radionic machine. In analog reality both remedy and patient frequency are individualized. Therefore the radionic dowsed specimen must be from the same plant as the flute or tincture to be accurately and predictably correlated.

EXAMPLE: Conditioned, exaggerated reactions of the immune system to foreign influences are called allergies. Western medicine, of course, believes that this can be explained entirely in terms of the lower dimensional, physical properties (chemistry) of the allergen.

Their typical binomial response is also an exaggerated reaction. The immune system is negated or completely turned off. They have a very poor record in dealing with allergies, as you well know, and turning off the immune system is also a far more dangerous risk to the patient than the allergy.

Allergy and conventional medicine's typical reaction to it are both pathological examples of the binomial mode processing we call an "automatic conditioned reaction." In binomial mode we have either no response or a triggered, full blown reaction with no options in between. The scenario goes something like this:

Binomial sees → Labels → Shuts down present tense sensory input → Reacts the same way it always did in the past. Unmonitored automatic <u>reactions</u> are binomial.

The immune system is, in fact, overreacting to the higher dimensional <u>energy</u> pattern of the offending substance. The solution obviously is to deactivate the trigger, and reprogram the immune system into a more appropriate analog mode response. Analog <u>responses</u> are based on present tense sensory inputs and are therefore more discriminating.

A full blown allergic reaction can be safely precipitated by experiencing the sound energy of an allergen. This response can then be meditated (adjusted) with acupuncture or Qi Gong treatments while the patient continues to listen to the offending sound. The body simply learns, by your example, how to produce a more graded and appropriate response. Exposures to that same substance in the future will result in a more reasonable, effective, and generally symptomless response.

EXAMPLE: Cancer is another condition that obviously originates in the implicated energetic profile of the patient. Here again the accepted, conventional materialist approach to treatment is superficial, inappropriate, and grossly counterproductive to good health. That approach is a serious problem for no other reason than it diverts attention, time, energy, and valuable resources away from the actual energetic problem and its resolution. I often have to spend more time treating the negative consequences of the conventional treatments than I do on the imbalance that manifested the cancer in the first place.

Analog medical techniques are very effective at dealing with this problem. I treat cancerous conditions with combinations of acupuncture, homeopathy, sound, and nutraceutics. The problem is always to keep your attention exclusively on the patient's energy profile. Remember, you can not send negating messages in the analog healing mode. **You will actually energize a wart or tumor by giving it your undivided attention**. In keeping with my tendency to contradict myself, I often make and use homeopathic remedies of their own solid tumors. In homeopathy this is called a "nosode" remedy.

Some degree of mental and life style reprogramming is very helpful if not essential in these cases. For human patients a change in diet, occupation, and prioritizing can often be critical. Practicing Tai Chi, meditation, and positive visualization techniques empower the patients to deal with their own condition. Doctor Zhi Gang Sha goes so far as to suggest **cancer patients can use the energy released from their own block to treat others**. (Absolutely brilliant) *XVI-27

For the animal patient, we must positively reinforce that all important bond with their owner. Training tips to alter behavior problems, instructions on massage and other energetic therapies the owner can employ at home are very helpful in this regard. You need to challenge the conventional fatalistic expectations commonly associated with this disease by showing them success stories. Pets will live out their owner's expectations, so give them positive ones. Also the owners need to be informed of the problems associated with commercial diets and conventional vaccination regimes.

EXAMPLE: I routinely reinforce or backup my analog energetic manipulations with generic binomial dietary adjustments and/or neutraceutic supplementation. I feel that you must supply the basic raw materials or building blocks the organizational plan needs to properly restructure the body.

The use of material supplementation appears to fly in the face of the principles of analog medicine. However it has some important potential energetic applications in addition to its material contributions. It is actually a more holistic approach.

Symbol and ritual are important and powerful elements in medicine. A change in the diet symbolizes a change to a healthy life style in the mind of our patients' owners. The nurturing feeding ritual helps reinforce their bonding by giving the owners something significant and important to do. Rituals also automatically predicate the expectation of positive results. They are essentially physically acted out positive affirmations or placebos. This type of positive reinforcement and support from the owner is much more important in the long run than mine. *XVI-18

I stress and frequently remind owners that all remedies and supplements should be given and or taken <u>mindfully</u>. In the objective materialistic world today supplements and medications are usually taken absent mindedly. We just assume they will work the way they are suppose to without us getting involved. But guess what? The patient is always involved, like it or not. Research has documented the power of the placebo effect. Placebo is another word for belief and focus. Research has also proved that grace improves digestion and utilization of food. Grace is mindfulness and intent. Therefore if we combine the effects of a proven physical supplement with mindfulness, belief, and intent, the results can and are frequently phenomenal or miraculous. * VI 5, XVI 15

This can also be called positive visualization, cyber-physiology, or grace for the spiritual or religious clients. When taking or giving an antioxidant, for example, you might imagine it as "Pac Men" eating up the free radicals or cancer cells. Calcium can be visualized as building up bone. MSM could be imagined carrying supplies or dissolving barriers. Prayers need not be so specific. *XVI-4, 11, 20

EXAMPLE: I often contradict my own advice and find myself treating my own or someone else's symptom rather than doing a complete balance. The secret to successful symptom resolution is essentially the same: analog mode. Symptoms are basically binomial with a name and emotional elements attached. Names come from past tense and include an expectation of consequence. So the first step is to reframe the symptom in strictly energetic and analog terms. The patient must focus from the mind on the character of the particular energy problem in the present tense. We must deal solely with the present tense energy block or imbalance rather than a perceived past tense cause. For example: Is the energy sharp or dull? Is it located or moving? This simple act, when properly done, shifts us right into the analog mode. The experience is immediately and dramatically changed. For example, the intensity of a painful situation will suddenly moderate. A great deal of the negativity of any experience is associated with anxiety, expectations, and a feeling of powerlessness (All binomial).

The next step is to engage the mind in actively correcting the unnamed energy problem with some type of appropriate manipulation or visualization. I usually move the energy down stream along the meridian pathways that are involved. This is done with Qi Gong, acupressure or suggestion. Pressure in the chest would be moved out and down the inside of the arm along the Heart, Lung, or Pericardium meridians, for example. You might choose to work with the particular chakra involved instead.

A brainful symptom reframed as a mindful energy block is manageable with analog technologies. An asthma attack becomes a dull pressure in the chest. A migraine becomes a sharp pain in the temples. Sciatica becomes a tingling or shooting pains down the leg. Anxiety becomes a dull knot in the stomach. We can only treat the energy block or imbalance with analog medicine, not the labeled condition or disease. Fortunately the condition or disease usually disappears after the energy is properly adjusted.

QUALIFICATION: Health is balance and that also applies to the modes of medical applications. Analog medicine is a present tense adjustment. This must be balanced with the practical cause and effects of the binomial mode. Many medical conditions are best attended to in basic binomial mode. It would not be practical or wise to initially deal with conditions such as obesity, tooth plaque, impacted anal glands, or fleas with analog therapies. Newtonian physics is also usually the logic of choice in the E.R.

In many cases analog methods can work but are impractical under the circumstances. I can replace displaced abomasums in cattle, in time, with analog methods. In most cases, however, the cow's diet can not be individually controlled, and so the displacement usually reoccurs. I can satisfactorily replace and physically secure it, in less than an hour, with surgery.

SUMMARY

Analog medicine is an amalgamation of the truly analog elements I found in the different alternative modalities that I personally studied. As such, it is less than completely comprehensive. It is however, a workable system that attempts to deal with the healing phenomena logically and therefore scientifically. It is a general philosophical approach or method rather than a dogma to be memorized and copied. Analog medicine is not about what to do, as much as how to do it. In quantum reality where healing originates, the individuality of the doctor cannot be separated from the treatment or the treated. It is only logical that the analog medical protocol you design and successfully employ will differ from mine.

SOUND THERAPY

To heal ourselves, or assist others to heal, we must focus in analog mode and stay there. To assume analog mode we select qualities or opt for attributes in the right hand column of the dichotomy list like analog, yin, and female.

These columns represent the extremes of any particular quality. The Chinese theory of yin/yang says that very few things, if any, are entirely yin or yang. It is often a matter of selecting a position that is more yin or analog than another.

Binomial-yang — Beta — Alpha — Theta — Delta —Analog-yin

Focus is accomplished by way of one or more of the brain's sensory organs. We can look, hear, smell, taste or feel. Individuals usually prioritize and use one sense more than the others, giving that sensation more weight. Their intuitive abilities and the language they use often reflect this prejudice.

A people who primarily focuses visually will dream and intuit in vivid pictures. They will say, "I see" when they mean, "understand."

Others may focus more on touch. Intuitive information comes to them as "just a feeling." They like to "keep in touch." "To get a feeling for," means to understand, for these people.

Those inclined to focus on sound will say that their intuitive information comes from a voice in their head. They want to tune in to your wave length, get tune-ups, and they like everything to be in harmony. When they understand, they will say, "I hear you."

Sound is the ears' interpretation of vibrations occurring in air or water. The mechanical transfer of movement through the media of matter causes the vibrations. On the physical level, this appears to be an example of the Q.M. principle of connectedness.

Sound waves are relatively long and compare to the "Extremely Low Frequencies" classification of the electromagnetic spectrum. The vital energy we refer to in regard to life forms has a microscopic length in comparison and would occur at the other end of the spectrum along with X-rays and Gamma rays. This disparity in size means that sound waves cannot effectively resonate with or directly affect the vital energy. On the other hand, huge sound waves are excellent carrier waves for the tiny vital energy waves of the vibrating element creating the sound. (analogy)

Spectra photometry is a science based upon this same principle but in regard to light waves. The light emitted by a combusted material carries the material's energy signature. In the same way, you cannot separate the light from the light source, you cannot separate the sound from the sounder.

If we decide to focus with the ear, sound will be the medium of transfer. Language and music are the disciplines involved.

LANGUAGE: Language can be used in either analog or binomial modes. In everyday use, the two modes are often mixed together. We know that abstract words and syntax are associated with extreme binomial mode. But language is also often used in the analog mode of "growl and purr." In this case, the words and phrases are like carrier waves to vent emotions, and the meanings of words and syntax is basically irrelevant. Profanity is of course an obvious example of analog growling. While the written word can be almost totally binomial, the transfer of strictly rational binomial information with spoken language is seldom achieved. Intent is an ever-present subliminal message in the voice and that is transferred in analog mode.

Vocalizations that do not involve abstract words like crying and laughing are mostly analog. Chants that involve nonsense "words" are fairly analog. Words that refer to non-material spiritual concepts like God, peace, and love are less binomial and more analog. Sentences that are memorized and repeated without cognition, such as prayers and blessings, are more analog.

People who meditate have established that intoning simple vowel sounds or repeating a mantra or a short prayer enhances their ability to maintain delta wave consciousness. (Delta = Analog)

MUSIC: Music is defined as the art and science of combining vocal or instrumental sounds. We can place the elements of music on a relative analog-binomial scale.

Analog—sound—pitch—beat—notes—melody—lyrics—Binomial

Music is not unlike language; the notes and melody are very similar to words and syntax. The lyrics are of course just that. All three elements are strongly binomial.

The beat is binomial because it breaks time into units. A steady beat that approximates the heartbeat is associated more with analog states. Regular steady beats have a boring quality that the binomial central nervous system eventually tries to ignore. Any irregular beat excites the central nervous system, inviting you to tap a toe or move, and therefore is more binomial.

High-pitched sound like the beta brain waves are more binomial, and the lower- pitched sound like the delta brain wave is more analog.

If we are concerned with healing and our sensory focus is by way of the ear, then elemental sound offers the most potential. This fact has not gone un-noticed. In ancient China, instruments were grouped into eight categories according to the material from which they were made. We group them today, in the western world, according to how they work: Strings, woodwinds, and percussion for example. *XVII-1

In Tibet, resonating bowls are made to balance the vital energy of the body. These bowls are made up of seven different metals, each from a different line in the periodic table of elements. The sound produced by these bowls when struck or rubbed carries the subtle energy of each of the metals to the listener. The note sounded by the bowl or bell is not as relevant to the energetic balance as the material it is made of.

Some of the stringed instruments made by Antonio Stradivari are considered to be the best ever made. All attempts to reproduce them have failed. Even computer-generated models that are exact physical replicas do not produce the same quality of sound. Subsequent research found that, for a full decade, he made all his instruments from the same tree. This was the same time period he produced his best work. Listeners are responding to the vital energy or timbre of that particular tree. *XVII-6

FLUTES: When I began doing research into the correlation of musical notes to the five elements of Chinese philosophy, I hoped to find a way to select acupuncture points based upon their resonance qualities. I was struck by the similarity of tuning or playing a musical instrument and selecting treatment points in acupuncture. What I needed was a way to expose patients in the field to a single musical note for a period of time while I attempted to divine resonate points. I found a local flute maker named Eric Hash, who made Lakota style flutes from various types of wood. I bought two because I really liked the different woods and the workmanship. I had no musical background, so Eric showed me the basic fingering pattern, and to my amazement the flutes just seemed to play themselves. I still believe my original idea of divining points is a good one, but it never proved to be very effective in my hands. I was unable to play the flute and look for points at the same time.

However while explaining my ideas to a group of people, another aspect became obvious. I noticed that when I started to blow on the cottonwood flute, a number of the people in the audience all simultaneously reached up to touch their upper chest area. From my perspective in front, it resembled synchronized swimming. I noticed the same reaction at a subsequent demonstration. It happened the instant I started to play and did not depend on which notes were used. I also found that I did not get this same reaction with the black walnut flute.

I got two more flutes from Eric and started to play around with the four different woods. People whom I had been using as energy "sensitives," or clairsentients, in most cases said that they could feel the vibrations in different parts of their bodies when I played each of the flutes in turn. The vibrations seemed to correspond to classic Chinese acupuncture meridians, and in many cases, we could actually see tiny muscle fasciculation along these pathways. People with no previous knowledge of meridians were able to accurately describe their courses while listening to the sound of a flute. Using blind trials on many different individuals, the results were amazingly consistent. In some individual cases movements were not perceived in the most common meridian reported but in another one that would be used to balance it in Chinese philosophy.

Not only were people identifying which meridians resonated with each flute, they were reporting in most cases the exact ones the Chinese thought were associated with that same plant as an herb. The choke cherry flute consistently resonates with the lung and large intestine meridians. The black walnut flute resonates with the kidney meridian. The red osier dogwood flute is consistently felt in the feet and legs because it resonates with the Dai or belt meridian. The hawthorn flute sound is usually felt specifically in the lower chest and heart area because it is associated with the pericardium, heart, spleen, and stomach meridians.

I have also found several excellent remedies that are not common knowledge. Antelope bitterbrush has a powerful effect on the neck, and Mountain mahogany on the prostate and spinal column, although I do not know which meridians are specifically involved yet. My Aspen has had the surprisingly consistent effect of resolving a congested sinus condition. This makes sense when you realize that it resonates with the bladder and stomach meridians, both of which drain the front of the face. None of my references list Aspen as a treatment for that condition.

I am continuing to collect different wooden flutes and have over eighty at last count.

I use the flutes for both diagnosis and treatments. The correct flute played for an asymptomatic patient will produce a classic release when an energy block is shaken loose. This is a very quick response that requires only a few notes to be played. Patients will say that they really like the sound, and it makes them sleepy "yawn." Patients who have allergies to a particular plant will experience a full-blown attack upon hearing the sound of that flute. We have several double blind instances to prove this point. Patients will not like the sound of flutes that resonate with meridians in which they already have excessive energy. Any flute that would aggravate their condition will irritate them, and in some cases result in a symptom such as a headache. So you can use both the positive and negative responses to the sounds to determine how the patient is unbalanced.

Pain is perhaps the easiest symptom to deal with because the patient is usually focused on it. If the patient complains of a headache, simply play the flute that resonates with the meridian involved. For example, aspen relieves a frontal headache, walnut works for the occipital one, and juniper relieves pain in the parietal area. The headache disappears instantly when the patient hears the sound of the appropriate flute.

You can apply the flute sounds using the same basic logic that you would use to prescribe herbs or homeopathic remedies for oral use. However, the results are experienced much faster, almost instantly in fact.

The effect is not lost in recording, so you can record the sound for use in the patient's personal meditation or healing sessions. Migraine sufferers, for example, might want to keep a walk-man handy with their specific flute sounds recorded in it.

I like to use the flutes for distant healing sessions. While I focus on the patients' picture or memory, I play the appropriate flute for them. The flute sound takes the place of visualizations or prayers that are traditionally used. In this case, the subtle energy (message) of the flute is carried by way of my focus to the patient rather than by the sound waves. This technique is most effective when the patient is asleep.

A flute is basically an analog instrument, and should be played intuitively with a continuous flowing style for the best results. Play it like you sign your name, with no abrupt changes between the notes. Do not focus on the notes, worry about how it sounds, or try to remember a particular song. Keep the patient in mental focus at all times and compose the music for them personally each time. If you don't feel comfortable playing music, just blow one monotonous note, concentrating on how the wood vibrates in your hands.

BOWLS: I also use vibrating (or singing) bowls for sound therapy. The bowls produce a monotone sound when struck or rubbed, and that sound transmits the subtle energy of the material it is made of. I use mostly the quartz bowls produced in this country as a spin off of the computer industry. Quartz is an excellent homeopathic remedy, in itself, which many patients can benefit from, but I use their sound mostly to carry the subtle energy of remedies for which I have no flute. I suspend a fresh plant, a tincture, or a homeopathic remedy inside the ringing bowl. The patient receives the subtle energy of the substance inside the bowel as well as the quartz energy via the sound wave. If you use plants that you have flutes made of, the two methods compare favorably. Clairsentients will usually report similar if not identical responses given the addition of the quartz factor. Again, I use the bowls to both diagnose and treat.

I have found these bowls to be particularly helpful on field trips. Sounding a live herb enhances its subtle message for the student. Individuals that normally do not consciously pick up on the energy of living plants will often experience something when the herb is placed in a ringing bowl. For example, many people report that the sound of Echinacea vibrates their upper sternum. This is, of course, the location of the thymus gland, which is usually associated with the immune system.

Allergic responses to the sound of a plant are similar if not identical to those produced by actual physical contacts. Those allergies can be safely, effectively, and permanently resolved in many cases if the symptoms can be suppressed with acupuncture or Qi Gong treatments while the patient continues to listen to the offending sound. The treatment is concluded when the patient reports no longer being irritated by the sound. This would be a variation of the N.E.A.T. protocol. *XVII-5

Sound therapy is a powerful tool that can be effectively combined with other forms of treatment. It is particularly dramatic when applied in a sweat lodge-like setting, taking advantage of group support and ritual.

Once you know which energy remedy that you want to use, it can be applied in several different ways. The more types of sensory input used, the more intense the experience.

1. The sound of a flute or bowl can be used for the auditory input.

2. Staring at the flames of an open fire assesses the subtle energy of the burning material. It is carried on the light waves to the eye. Notice how everyone gazes into the fire on a camping trip. It just isn't camping unless you have a wood fire. Next time you go camping, pay particular attention to the quality of the flames and how they affect you as you burn different woods.

3. The heat from the fire also carries this subtle energy. That is the idea behind moxibustion in Chinese medicine. I am saying that you don't have to limit yourself to the herb mugwort, in other words.

4. The scent of a freshly picked herb or the smoke from burning dried herbs also carries these energies to the nose. These can both be effectively used in treatments. Amerindians commonly use smudging, Aromatherapy has been well documented, and the East Indians have used incense for centuries for similar purposes.

5. The subtle energy is of course carried in the physical material of the plant. So the patient can orally consume it. Herbal

remedies are well known. When the remedy is chemically poisonous, the energy can be transferred to water or alcohol as a homeopathic.

6. The live plant is the most potent source of unadulterated vital energy patterns, and the simple act of focusing attention on a living organism accesses this energy. You need to focus in a meditative or prayerful state, consciously (mindfully) experiencing the energy in every way possible. Everyone enjoys a walk in the garden or woods where live plants are growing. Houseplants are essentials for most people. You can grow your remedy as a houseplant and use it as a meditation devise rather than physically consuming it or making a flute out of it.

Processing a remedy destroys or alters its vital energy. A cooked, microwaved, or refined plant remedy is much less vital than the original raw material or live plant.

JUST IMAGINE

Imagine taking a patient in need of a piñon pine remedy out into the Utah desert on a starry night. Set up camp under an old piñon, start a fire of piñon wood, gather some piñon nuts and play a tune on your piñon flute. Would that be magic or what?

I have been amazed by the abilities of patients to intuit their own tree remedy. When I first tell them that we are going to try and find a tree remedy for them, a fair number of them will offer a suggestion as to what it might be. They are usually right.

SUMMARY

Sound therapy differs from and is not music therapy. In music therapy the focus is from the binomial brain and its auditory sensor, the ear. The sensory information it receives (sound) is censored and interpreted into meaning. Music therapy affects the patient on the lower implicated dimensions of emotion and rationality. In sound therapy the patient focuses from the perspective of the analog mind on the subtle energy profile of the vibrating plant material. The energy is experienced on the higher implicated dimensions of consciousness.

Sound therapy is based upon the "entanglement principle" of Quantum Mechanics. The sound cannot be separated from the sounder or vibrator. The elemental sound of a vibrating material is used to treat a patient's energy body or balance in the same way herbal remedies would be used in classic Chinese medicine. The sound of an herbal remedy is a powerful and rapid way to effect an energetic adjustment. As with Homeopathic remedies, chemically poisonous materials can be utilized without risk.

CHRIST TAUGHT QUANTUM MECHANICS

Jesus was primarily concerned with salvation, redemption, love, and healing. Above all else, he was a teacher. His teachings have always seemed to be at odds with the conventional picture of reality painted by Newtonian physics. The things that He said and did were called "miracles" because they could not be explained in terms of that logic. His description of reality appears to correspond to the one explained in Quantum Mechanics. Viewed from that perspective, Christ's message appears to be perfectly logical. *XIII 4

Salvation and redemption share a common quality with healing. All three deal with the recovery of a previously healthy or preferred state of being. This idea is covered by the word "*remanifestation*" in my definition of healing. Healing comes from a higher implicated dimension of the patient, and it is fueled by his own vital energy. Healing always involves a learning process for the patient. Healing is learning to be healthy. Christ was teaching man how to choose or learn the healthy state of being. Resisting the basic laws of nature creates tension (disease) in the life process. A process that follows or unifies with these laws is more efficient and contains less tension or

327

stress. Health is the harmonious and easiest way of being. A non-judgmental, accepting unity is called love.

HEALING or SALVATION = LEARNING = LOVE = HARMONY

= UNITY

The principles that Jesus taught were simple, and he repeated them again and again. They describe a preferred state of being that is an analog quantum reality.

ONE GOD: Christ taught that there was only one true God. The spiritual reality that He called God can't be separated into parts. It is all-inclusive, everywhere, timeless and connected. This is also the definition of analog reality and the quantum principle of connectedness or entanglement. The classic definition of God is indistinguishable from the definition of the quantum field.

Here, I must admit, He deviates from quantum logic in giving God the male gender or any gender at all for that matter. I believe that He was describing the basic principle of connectedness, unity, and love by way of analogy to the parent-child relationship.

LOVE: Once again we have a reference to the principle of connectedness or unity. Love defines a non-judgmental acceptance.

There is identification with common interests and sharing.

FAITH-BELIEF: These are the anti-thesis of rational understanding. We say we understand when a cause and effect relationship is perceived. Faith is a rejection of that concept and the binomial mode where it manifests. Atomic physicists believed that a specific sub-atomic "particle" existed before they could actually demonstrate it. The act of believing in or focusing on one of the infinite number of possibilities increases the probability that it will ultimately manifest for us. Quantum scientists would say belief collapses the state vector to a single wave form.

NON-JUDGEMENT: Christ repeatedly admonished us for judging or categorizing others or ourselves. Categorization breaks mankind into subgroups such as good and bad. This is the definition and essence of the binomial mode. According to the chapter on processing, labeling or naming automatically blocks further sensory input and information exchange. It breaks the connection we define as unity, love, and spirituality.

FORGIVENESS: Forgiveness releases vital energy locked up in the memory. Healing is generated by the timeless analog mode.

Concentrating the vital energy, in the present tense, by way of focus, maximizes healing.

The GOLDEN RULE: This commandment attributed to Christ is another statement of unity and connectedness. By doing unto others as we would have them do unto us, we are accepting a unity of purpose with the rest of mankind. We are identifying with mankind as a whole, as opposed to our separate egos, family, or group. He also made it a positive statement rather than a negating one. Analog mode is unable to send or receive negatives.

LOVE THY ENEMY: Progress in spiritual enlightenment requires learning specific lessons. Experiencing explicated reality is how we do this. One of the lessons to be learned in physical reality is forgiveness. Someone must transgress against you before you can forgive them. Without the transgressor the lesson can't be learned. Another important spiritual lesson is empathy. You must experience pain, grief, hatred, lust, jealousy, etc. in order to truly empathize with another person. In this context, those things typically considered or judged bad are reframed as opportunities to make these important steps up into the higher dimensions of spirituality.

PRAYER: The description of prayer He gave in "The Sermon on the Mount" provides some excellent guidelines for accessing the higher implicated dimensions, and it closely parallels those suggested by meditators. They both attempt to focus our conscious energy on the implicated dimensions by selecting analog qualities and mechanisms.

Praying in a closet blocks out our brain's binomial sensory organs, allowing us to focus inward and upward or from the mind's perspective. The use of earphones and blindfolds by modern meditators achieves the same effect.

Praying in a closet also effectively eliminates the potential for the binomial activities of doing and going. The traditional prayer position essentially achieves this same passive state by incapacitating the hands and legs.

The closet also insulates us from the social contacts ("I hope that I don't look stupid.") that can interfere with our focused intent.

We normally have a lot of trouble letting go of physical reality and the thoughts that accompany it. When we focus on material reality, our brain wave pattern is characterized by high frequency, low amplitude waves. Learning to meditate or pray effectively means that we learn to focus in a way that produces progressively lower frequency, higher amplitude brain waves. Any thought that includes references to physical reality and cause and effect will automatically generate higher frequency waves and a left shift into binomial mode. "The Lord's Prayer" is carefully and purposely worded to avoid any references to the binomial qualities in the left column of our dichotomy list.

"<u>Our</u> <u>Father</u> <u>Who</u> <u>art</u> <u>in</u> <u>heaven</u>." Father is a reference to our connectedness and identity with the implicated codes of the quantum field. We are physically connected by way of genetics to our parents and children, and that relationship should be one of non-judgmental, loving acceptance. By analogy we should consider our relationship to the implicated code (laws of nature) in the same way. Heaven is analog reality and is not physically located in binomial space-time.

<u>Hallowed</u> <u>be</u> <u>Thy</u> <u>name</u>." The name is analog individualization as opposed to a binomial labeling and categorization: God, rather than "a god." Focus on "Bob" instead of "the patient."

"<u>Thy</u> <u>kingdom</u> <u>come</u>." The laws of nature, the encoded plans, souls, the Great Spirit and/or God are of the implicated dimensions. They are timeless, unchanging, and will eventually manifest given enough time in explicate reality.

"Thy will be done." "On earth as it is in heaven." A clear statement describing the manifesting of implicated codes, laws, or plans in the explicated lower dimensions.

"Give us this day our daily bread." Again he rejects the idea of cause and effect. He is describing a passive state of being.

"Forgive us our debts as we forgive our debtors." Live or focus in the present tense to maximize the vital energy needed to fuel the healing process. Energy invested in historic grudges and self-condemnations is unavailable for healing.

"And lead us not into temptation," The sensory functions that lead to temptations are binomial. When we are physically tempted we are focused on the lower explicated dimensions. To heal we must focus on the higher implicated dimensions.

"But deliver us from evil." Evil is morally bad or wrong conduct. A moral judgment is of the binomial mode and the lower dimensions. Evil is also considered the cause of bad. Cause and effect is not evident in the higher spiritual, dimensions. The demons of life are manifestations of the tensions we create with our judgments and rejection of unity. They are transformed by understanding and acceptance. Scientists today call this the entanglement principle. Christ called it love.

"For Thine is the kingdom, the power and the glory forever." Implicated reality contains more potential energy than the materialized ones and it is timeless. This phrase was apparently not part of the original prayer.

When I was originally introduced to the teachings of Jesus Christ, I was a young boy in Iowa. I got the distinct impression that He was bigger than life. He did and said things that were beyond our comprehension and understanding. Religion was a matter of blindly (ritualistically) following some rules and not doing a lot of things.

After I started studying healing, some things began to fall into place, and I was able to frame Him in a different way. The words attributed to Christ could be taken quite literally. He was, in fact, describing a reality of healing that modern science is just now beginning to understand and delineate as Quantum Mechanics.

SUMMARY

According to the historic records, Christ did not reserve the miracle of healing for himself. He taught the disciples to use healing as part of their preaching. He was clear on the fact that anyone who followed His instruction could learn to heal themselves and others. In so many words, He said that we must access the analog mode and process with the principles and logic that modern man now calls Quantum Mechanics.

IMPORTANT PRINCIPLES I DISCOVERED IN THE ALTERNATIVE MEDICAL PROTOCOLS

I learned some important lessons and valuable quantum technologies in each of the alternative protocols that I studied. Some examples follow:

Acupuncture:

There are many correct ways to practice a protocol. Every master and school of acupuncture teach it differently. This fact was not appreciated by the first Western translators, and so much of our original exposure to this discipline was a confusing jumbled mix of incompatible and contradicting philosophies and approaches.

Disease can be analyzed and dealt with entirely as an energetic problem. Mindful solutions are generally only effective if the problem can be framed in terms of energy. The Chinese theories concerning circulating Chi, the five elements, and yin/yang allow for this perspective.

Radionics:

Most people will need some sort of ritual and physical devises to effectively evaluate and manipulate subtle energy. Intuitive information is vague by definition and does not lend itself readily to analysis. Recording devices, such as radionic machines, can provide a mechanism for sharpening our focus, expanding our attention span, organizing the intuitive effort, and storing pertinent information.

Homeopathy:

The medical solution to physical insult is often the mirror imaged subtle energy profile of the offending material. The energy of a potent poison is potentially an effective remedy for the symptoms produced by its physical-chemical form.

The more completely we disassociate the subtle energy from its material base, the stronger the healing effect. Potency is a direct measure of this principle, and the higher potency remedies are more dramatically effective.

The more specific a remedy is the more effective it is. The highest potency remedies have the narrowest range of patient effectiveness.

Some effects of specific disease conditions can be propagated to the next generation. The homoeopathists call these phenomena miasms, while modern physical scientists call it genetic transference.

Healing is not always pleasant. A healing crisis or detoxification can follow an appropriate and successful treatment. Both are positive indicators, but they are unpleasant ordeals.

Rieki:

You do not need any physical gadgets to do effective healing work. Rieki practitioners often get impressive results using only their consciousness, intent and visualizations.

Zero balancing:

You do not need to move or use force to produce good results. The fulcrum may be the most powerful tool and it does not move. Placement is the only adjustment required.

Meditation:

The key to successful healing is to slow down the basic brain wave pattern. People who meditate simply learn practical ways to progressively reduce the frequency of their brain wave activity.

Many disease conditions are produced by our own inappropriate mental programming. Meditation teaches us to disassociate self limiting and disease promoting mental programs by denying them vital energy input.

Western Herbalism:

Indigenous plants are more effective. Local circumstances provide for a type of resonance or connection between all living organisms in the eco-system. For example, Amerindians typically get much better results with North American herbs than Afro-Americans do.

Plants have individualized profiles. Many factors can contribute to profile differences in the plants of a given species and even in a growing population. Rolling Thunder always picks very specific plants for a patient's remedy.

Kinesiology or Muscle Testing:

Subtle energy is not significantly constrained by typical containers. These tests are routinely and successfully performed on products sealed in containers.

Most people have the ability to access subtle energy information. Muscle testing is widely used to assess the compatibility of substances on a wide array of people and animals.

Shamanism:

Even the thought of doing something compromises the healing mind set. Shamans typically enlist the help of a power or spirit animal to do the imagined physical manipulations of the healing ritual. This allows them to remain totally passive.

Social connection is an important element in healing. The sweat lodge healing ceremony enlists the help and support of the whole community in the effort.

Prayer:

Belief and intent are more important than understanding. All the studies indicate that the "Thy will be done" prayer is the most effective form. Knowledge of the details involved in the healing process is not required. In some cases focusing on the perceived details actually interferes with the process by distracting our focus from the intent.

Qi Gong:

Visualization is all that is ultimately necessary to move energy. Some idea of normal energy flow is useful in achieving balance.

To become effective healers we have to discipline ourselves mentally and physically with regular practice. Qi Gong practitioners emphasize that we first need to mantain our own balance with daily practices if we want to be effective healers.

SILVA:

Self hypnosis is easy, usable and a valuable tool. The Silva method has many practical techniques that you can use to facilitate the shift to the right and the higher states of consciousness.

BACH FLOWERS:

The sun's radiation will effectively pickup and transfer implicated energy information. This mechanism is used, in this protocol, to transfer a flower's subtle energy profile to water. We can use this same principle to help transfer our implicated healing intent to a patient. In other words, always keep the sun on your back and try to position the patient in your shadow.

ALTERNATIVE MEDICNE IN GENERAL:

The life process is more than its material manifestation.

Practicing medicine is more than selling the latest patented drugs and gadgets.

RECOMMENDED READING and BIBLIOGRAPHY

I Introduction

1 *Why People Don't Heal and How They Can*
   Caroline Myss   1997   Harmony Books

2 *Rolling Thunder*
   Doug Boyd   1974   Dell Publishing

3 *Acupuncture Point Combinations;* The Key to Clinical Success
   Jeremy Ross   1995   Churchill Livingstone

III Dimensions of Reality

1 *Hands of Light*; A Guide to Healing Through the Human Energy
   Field
   Barbara Brennan   1988   Bantam books

2 *The Holographic Universe*
   Michael Talbot   1991   Harper Collins

3 *The Elegant Universe*; Superstrings, hidden dimensions and the
   quest for the ultimate theory
   Brian Greene   1999   W. W. Norton & Company

4 *Science   and   Human   Transformation*;   Subtle   energies,
   intentionality and consciousness
   William A. Tiller   1997   Pavior

5 *Zero;* The biography of a dangerous idea
   Charles Seife   2000   Viking-Penquin Books

6   *The Medium the Mystic and the Physicist;* Toward a general
theory of the paranormal
Lawrence LaShan   1974   Penquin Books Ltd.

7   *Edgar Cayce's story of the origin and destiny of man*
Lytle Robinson   1976   Berkley

8   *The End Of Time;* The Next Revolution in Physics
Julian Barbour   1999   Oxford University Press

9   *Spiritual Healing*; Scientific Validation of a Healing Revolution
Daniel J. Benor, M.D.   1992   Vision Publications

IV Its Only Logical

1   *A World of Ideas*; A dictionary of important theories, concepts,
beliefs, and thinkers
Chris Rohmann   1999   Ballantine Books

2   *The Way of the Wizard*; Twenty Spiritual Lessons for Creating the
Life You Want
Deepak Chopra   1995   Harmony Books

3   *The Marriage of Sense and Soul*; Integrating Science and Religion
Ken Wilber   1998   Random House

4   *The Secret Life of Plants;* A fascinating account of the physical,
emotional and spiritual relations between plants and man
Peter Tompkins and Christopher Bird
                          1973   Perennial Library; Harper & Row

5   *The Secret Life of Your Cells*
Robert B. Stone   1989   Whitford Press

6   *The New Physics of Healing*; Inside the Medicine of the Future
Deepak Chopra   1990   Sounds True Audio

7 *The Physics of Consciousness;* The quantum mind and the meaning of life
Evan Harris Walker   2000   Perseus Publishing

V The Yin And Yang of Neurophysiology

1 *The Body Electric*; Electromagnetism and the Foundation of Life
Robert O. Becker and Gary Selden
                          1985   Quill; William Morrow

2 *Thinking in Pictures*; and other reports from my life with autism
Temple Grandin   1995   Doubleday

3 *The Web That Has No Weaver*; Understanding Chinese Medicine
Ted J. Kaptchuk   1983   Congdon & Weed Inc.

4 *The Foundations of Chinese Medicine*; A Comprehensive Text for Acupuncturists and Herbalists
Giovanni Maciocia   1989   Churchill Livingstone

5 Fundamentals *of Chinese Acupuncture*
Andrew Ellis, Nigel Wiseman, Ken Boss
                          1991   Paradigm

6 *Chinese Acupuncture and Moxibustion*
Cheng Xinnong   1987   Foreign Language Press

7 *Chasing the Dragon's Tail*
Yoshio Manaka, Kazuko Itaya and Steven Birch
                          1995   Paradigm Publications

8 *Left Brain; Right Brain*
Sally P. Springer and George Deutsch
                          1993   W.H. Freeman and Company

9   *The Emerging Mind*
Karen Nesbitt Shanor   1999   Renaissance Books

10 *A User's Guide to the Brain*; Perception, Attention, and the Four Theaters of the Brain
John J. Ratey   2001   Random House

11 *The Astonishing Hypothesis*; The Scientific Search for the Soul
Francis Crick   1994   Touchstone

12 *The Large, The Small and the Human Mind*
Roger Penrose   1997   Cambridge Un. Press

13 *The Dragons of Eden*; Speculations on the Evolution of Human Intelligence
Carl Sagan   1997   Ballantine Books

VI Processing

1   *The Law of Psychic Phenomena*
Thomson J. Hudson   1995   Castle Books

2   *Drawings on the Right Side of the Brain*
Betty Edwards   1979   Jeremy P.Tarcher

3   *Hypnosis for Change*
Josie Hadley and Carol Staudacher
                    1996   New Harbinger Publications

4   Hypnosis, *Acupuncture and Pain*
Maurice M. Tinterow   1989   Bio Communications Press

5   *The Placebo Response;* How You Can Release the Body's Inner
    Pharmacy for better Health
    Howard Brody with Daralyn Brody
                        2000   Cliff   Street   Books;
                                        Harper Collins

                        VII Thinking

1   *Language Thought and Reality*; Selected writings of Benjamin
    Lee Whorf
    John B. Carroll   1995   The Mit Press

2   *King Solomon's Ring;* New Light on Animal Ways
    Konrad Z. Lorenz   1952   Thomas Y. Crowell
                                         Company

3   Tom   Dorrance   Talks   About   Horses;   *True   Unity*;   Willing
    Communication Between Horses and Humans
    Tom Dorrance and Milly Hunt Porter
                        1987   Give It a Go Enterprises

4   *Horse Follow Closely*; Native American Horsemanship
    GaWani Pony Boy   1998   Bow Tie Press

5   Voices *of the First Day;* Awakening in the Aboriginal Dreamtime
    Robert Lawlor   1991   Inner Traditions
                                        International

6   *Indian Medicine Power*; Medicine people from numerous tribes
    demonstrate how ancient medical practices can be used to attain
    mind and body wisdom and to "walk in balance" with nature.
    Brad Steiger   1984   Schiffer Publishing;
                                        Whitford Press

7   *Fools Crow Wisdom and Power*; in dialogue with the great Sioux
    Holy Man, Fools Crow.
    Thomas E. Mails   1991   Council Oak Books,
                                               Tulsa

8   *Black Elk;* The Sacred Ways of a Lakota
    Wallace Black Elk and William S. Lyon
                             1990   Harper San Francisco

9   *The World is as You Dream It*; Shamanic Teachings From the
    Amazon and Andes
    John Perkins   1994   Destiny Books

10 *American Indian Healing Arts*; Herbs, Rituals, and Remedies for
    Every Season of Life
    E. Barrie Kavasch and Karen Baar
                             1999   Bantam Books

11 *The Secret Power Within*; Zen Solutions to Real Problems
    Chuck Norris   1996   Broadway Books

12 *Men are From Mars, Women Are From Venus*; A practical guide
    for improving communication and getting what you want in your
    relationships
    John Gray   1993   Harper Audio

13 *The Tao of Physics*: An Exploration of the Parallels between
    Modern Physics and Eastern Mysticism
    Fritjof Capra   1999   Shambhala

14 *Wholeness and the Implicte Order:*
    David Bohm   1980   Routledge

15 *THEPRESENCEOF THE PAST:* Morphic Resonance & The
    Habits of Nature
    Rupert Sheldrake   1995   Park Street Press

IX Movement-Change -Learning-Evolving

1   *A Brief History of Time*; From the Big Bang to Black Holes
    Stephen W. Hawking   1988   Bantam Books

2   *Dianetics;* The Modern Science of Mental Health
    L.Ron Hubbard   1992   Bridge Publications Inc.

3   *Meaning and Medicine*; Lessons from a Doctor's Tales of
    Breakthrough and Healing
    Larry Dossey   1991   Bantam Book

4   *Atom;* An Odessey from the Big Bang to Life on Earth...and
    Beyond
    Lawerence M. Krauss   2001   Little Brown and
                                           Company

5   *The Hole in the Universe*; How Scientists Peered Over the Edge of
    Emptiness and Found Everything
    K.C. Cole   2001   Harcourt

6   *The Dancing Wu Li Masters*; An Overveiw of the New Physics
    Gary Zukav   1980   Bantam Books

7   *The FIELD;* The quest for the secret force of the universe
    Lynne McTaggart   2002   Harper Collins

X The Limits of Binomial Logic

1   *The Puzzle of Pain*
    Ronald Melzack   1973   Basic Books Inc.

2   *Genome*; The Autobiography of a Species in 23 Chapters
    Matt Ridley   2000   Perennial;
                                            Harper Collins Pub.

XI Just Pretend

1 *Minding the Body Mending the Mind*
   Joan Borysenko   1987   Bantam Books

2 *The Power of the Mind to Heal;* Renewing Body, Mind and Spirit
   Joan Borysenko   Audio Nightingale Conant

3 *The Physician Within You;* Medicine for the Millennium
   Gladys Taylor McGarey with Jess Stearn
                            1997   Health Communications Inc.

4 *Meaning and Medicine*; Lessons From a Doctor's Tales of
   Breakthrough and Healing
   Larry Dossey   1992   Bantam

5 *Quantum Healing*; Exploring the Frontiers of Mind/ Body
   Medicine
   Deepak Chopra   1990   Bantam

6 *Mental Training for Peak Performance*
   Steven Ungerleider   1996   Rodale Press Inc.

XII The Discipline of Healing

1 *Dianetics*; The Modern Science of Mental Health
   L. Ron Hubbard   1992   Bridge Publications Inc.

2 *Timeless Healing;* The Power and Biology of Belief
   Herbert Benson with Marg Stark
                            1996   Simon and Schuster Inc.

3 *Emotional Alchemy;* How the Mind Can Heal the Heart
   Tara Bennett-Goleman   2001   Harmony Books

4 *The Management of Evolutionary Change*
   William Harris   1994   Centerpointe Research
                                          Institute

5 *Wherever You Go, There You Are;* Mindfulness Meditation in
   Everyday Life
   Jon Kabat-Zinn   1994   Hyperion

6 *Imagery In Healing*; Shamanism and Modern Medicine
   Jeanne Achterberg   1985   Shambhala Publications Inc.

7 *Breathing Under Water*; The Inner Life of Tai Chi Chuan
   Margaret Emerson   1993   North Atlantic Books

8 *Centering;* A Guide to Inner Growth
   Sanders G. Laurie and   1983   Destiny Books
   Melvin J. Tucker

9 *Tai Chi Chuan;* The Internal Tradition
   Ron Sieh   1992   North Atlantic Books

10 *Reiki;* Universal Life Energy
   Bodo J. Baginski and   1988   Life Rhythm
   Shalila Sharamon

11 *Awakening Intuition;* Using your Mind Body Network For Insight
   and Healing
   Mona Lisa Schulz   1990   Harper Row

12 *Your Sixth Sense*; Activating Your Psychic Potenial
   Belleruth Naparstek   1997   Harper Collins
                                          Publishing Inc.

13 *Meditation as Medicine;* Activate the Power of Your Natural
   Healing Force
   Dharma Singh Khalsa and Cameron Stauth
                           2001   Pocket Books

14 *The Secrets to Manifesting Your Destiny;* Audio
    Dr. Wayne W. Dyer   Nightingale-Conant

XIII Divining

1 *The Diviner's Handbook*; A Guide to the Timeless Art of
   Dowsing
   Tom Graves   1990   Destiny Books

2 *Radionics and Progressive Energies*
   Keith Mason   1984   C.W. Daniel
                                               Company Limited

3 *Radionics and the Subtle Anatomy of Man*
   David V. Tansley   1985   C.W. Daniel
                                               Company Limited

4 *Five Elements and Ten Stems*; Nan Ching Theory, Diagnostics
   and Practice
   Kiiko Matsumoto and Steven Birch
                            1983   Paradigm Publications

5 *Introduction to Meridian Therapy;* Japanese Classical
   Acupuncture
   Shudo Denmei translated by Steven Birch
                            1990   Eastland Press

XV Working With Energy

1 *Self Healing*; Powerful Techniques
   Ranjie N. Singh   1998   Health Psychology
                                               Associates Inc.

2   *Exploring the Spectrum*
    Philip S. Callahan   1994   Acres USA

3   *Color Medicine*; The secrets of color/vibrational healing
    Charles Klotsche   Light Technology Publishing

4   *The Book of Magnetic Healing and Treatments*
    Noel C. Norris   1995 International Research and
                               Development Magnetic Health
                               Products Organization

5   *Vibrational Medicine for the 21ˢᵗ Century*
    Richard Gerber   2000   Harper Collins

6   *The Healing Power of Color;* How to use color to improve your
    mental, physical and spiritual well-being
    Betty Wood   1984   Destiny Books

7   *Edgar Cayce on Healing*
    Mary Ellewn Carter and William A. McGarvey
                               1972   Warner Books

8   *Magnetic Therapy*; The Pain Cure Alternative
    Ron Lawrence and Paul J. Rosch and Judith Plowden
                               1998   Prima Health

9   *The Pain Relief Breakthrough*; The power of magnets to relieve
    backaches, arthritis, menstrual cramps, carpal tunnel syndrome,
    sports injuries, and more
    Julian Whitaker and Brenda Adderly
                               1998   Little, Brown and Company

XVI Analog Medicine

1   *Zero Balancing*; Touching the Energy of Bone
    John Hamwee   1999   Frances Lincoin limited

2 *Qi Gong Therapy*; The Chinese Art of Healing with Energy
   Tzu Kuo Shih   1994   Station Hill Press

3 *Medical Qi Gong;* Vital Energy Build-up Exercises - Volume 1
   Steven KH Aung   1996   World National
                                           Medicine Foundation

4 *You The Healer;* The World Famous Silva Method on How to
   Heal Yourself and Others
   Jos'e Silva and Robert B. Stone
                           1989   H.J. Kramer Inc.

5 *The Energetics of Western Herbs;* Treatment Strategies
   Integrating Western and Oriental Herbal Medicine
   Peter Holmes   1997   Snow Lotus Press

6 *Jade Remedies;* A Chinese Herbal Reference for the West
   Peter Holmes   1996   Snow Lotus Press

7 *Oriental Materia Medica*; A Concise Guide
   Hong-Yen Hsu and associates
                           1986   Keats Publishing Inc.

8 *Reinventing Medicine;* Beyond Mind/body to a New Era of
   Healing
   Larry Dossey   1999   Harper; San Francisco

9 *Therapeutic Touch*; the Theory and Practice of
   Jean Sayre-Adams and Steven Wright
                           1995   Churchill-Livingstone

10 *Accepting Your Power to Heal*; The Personal Practice of
   Therapeutic Touch
   Dolores Krieger   1993   Bear and Company

11 *Love, Medicine and Miracles*
   Bernie S. Siegel   1986   Harper and Row

12 *The Tellington-Jones Equine Awareness Method*; An introduction to Linda Tellington-Jones 1988   Breakthrough Publications Inc. and Ursula Bruns

13 *The Way of the Shaman*
    Michael Harner   1990   Harper-Collins

14 *Shaman Healer Sage*; How to Heal Yourself and Others With the Energy Medicine of the Americas
    Alberto Villoldo   2000   Harper-Collins

15 *Healing Words*; The Power of Prayer and the Practice of Medicine
    Larry Dossey   1993   Harper- Collins

16 *Bach Flower Therapy*; Theory and Practice
    Mechthild Scheffer   1988   Healing Arts Press

17 *Flower Essences*; Reordering Our Understanding and Approach to Illness and Health
    Machaelle Small Wright 1988   Perelandra, Ltd.

18 *The Power of Ritual*
    Rachel Pollack   2000   Dell Publishing

19 *New Bach Flower Body Maps;* Treatment by Topical Application
    Dietmar Kramer and Helmut Wild
                                 1996   Healing Arts Press

20 *Spiritual Healing*; Scientific Validation of a Healing Revolution
    Daniel J. Benor and   2001   Vision Publications
    Larry Dossey

21 *Health and Healing;* A Look at Medical Practices-from Herbal Remedies to Biotechnology-and What They Tell Us About
    Andrew Weil   1988   Houghton Mifflin Company

22 *The Nature of Animal Healing;* The Path to Your Pet's Health, Happiness, and Longevity
Martin Goldstein   1999   Alfred A. Knopf

23 *Miracle Healing from China QiGong*
Charles T. McGee with Effie Toyyewchow
                         1994   Meditress

24 *Organon of Medicine*; Translated by William Boericke
Samuel Hahnemann   1992   B. Jain Publishers [P] Ltd.

25 *Twelve and Twelve in Acupuncture*; Advanced Principles and Techniques
Richard Tan and Stephen Rush
                         1991   Self published

26 *Twenty-Four More in Acupuncture*; Advanced Principles and Techniques
Richard Tan and Stephen Rush
                         1994   Self published

27 *Power Healing*; The four Keys to Energizing Your Body, Mind and Spirit
Dr. Zhi Gang Sha   2002   Harper San Francisco

XVII Sound Therapy

1 *The Secret Power of Music*
David Tame   1984   Turnstone Press Ltd.

2 *Music Therapy for Non-musicians*
Ted Andrews   1997   Dragonhawk Publishing

3 *The Mozart Effect*; Tapping the Power of Music to Heal the Body, Strengthen the Mind and Unlock the Creative Spirit
Don Campbell   1997   Avon Books

4 *The Tao of Music;* Sound Psychology
   John M. Ortiz   1997   Samuel Weiser

5 *Veterinary NAET*: The Veterinary Application of NAET; A
   Breakthrough Approach to Allergy Resolution
   Roger W, Valentine and 1998   Journal of the American
   Jan K. Steele   Holistic Veterinary Medical
                                           Association Vol 17 No. 1

6 *Measure for Measure*; A musical history of science
   Thomas Levenson   1994   Simon and Schuster

XVIII Christ Taught Quantum Mechanics

1 *Bible*; King James

2 *The Complete Jesus;* All the sayings of Jesus gathered from
   ancient sources and compiled into a single volume for the first
   time Ricky Alan Mayotte   1997   Steerforth Press

3 *How to Know God;* The souls Journey Into the Mystery of
   Mysteries
   Deepak Chopra   2000   Harmony Books

4 *God at the Speed of Light;* The Melding of Science and
   Spirituality
   Lee Brumann M.D.   2001   A.R.E. Press

Ronald L. Hamm DVM

GLOSSARY

Acupuncture: The practice of stimulating very specific points on the surface of the body with the intent of adjusting the patient's energy profile and health. That stimulation can be accomplished in many ways. Pressure, needles, heat, herbs, magnets, lasers, light, and electricity have all been used successfully by practitioners.

Acupuncture points: Physically defined points on the surface of the body that provide direct access to the chi flow. They are used to both diagnose and treat problems in chi flow and distribution. They can be physically located with electrical instruments as areas of low resistance.

Affirmation: Vocalized positive statements designed to influence the subconscious analog mind. Used by many to overcome negative thought patterns, habits, and compulsive behaviors.

Ah Shi point: Tender points upon palpation. These are not necessarily one of the recognized acupuncture points. They are considered to be diagnostic by many practitioners.

<u>Alarm</u> <u>Points</u>: Diagnostic points on the front of the body that are correlated with problems in the different meridians. Pain upon palpation is diagnostic.

<u>Alpha</u> <u>Waves</u>: EEG brainwave frequencies in the 8 to13 hertz range. This state of mind is often associated with prayer, meditation, and intuitive perceptions. It is generally accepted as the mind set of "centering" or present tense focus. Many established healers display this level of activity when they are working on a patient.

<u>Amplitude</u>: The quantitative difference between the highest positive value and lowest negative value in a waveform. Amplitude determines the degree or intensity of effect a waveform will have in reality.

<u>Analog</u>: Connectedness, sameness, flowing, continuously varying, and spectral are all words that describe this quality. Cursive is analog writing.

<u>Aromatherapy</u>: A medical or religious practice utilizing the odor of different herbs and other aromatic substances. Most western practitioners of aromatherapy today use commercially prepared aromatic oils. The eastern traditions have traditionally burned solid

gum and resin preparations called incenses. The American Indian uses burning or smoldering herbs for this purpose and it is called smudging. The aerosolized substance is thought to deliver the therapeutic effect of the substance to the patient.

Aura: The visual appearance of some of the higher dimensions associated with the physical body. Theoretically we are all capable of seeing and feeling some aspects of this energetic reality but in our early development we learned to ignore it. Contrary to opinion, the aura is not given off by or projected out from the physical body. The aura projects down from above and part of it materializes or manifests as the body.

Bach Flower Remedy: Edward Bach made sun generated water remedies from the flowers of different plants to treat mental or psychological problems. He believed that most physical diseases originated in the personality profile of the patient.

Beta Waves: EEG brainwave activity in the 13 to 30 range. This high pitched activity is present during most wakeful conscious activity such as goal oriented physical activity, talking, and thinking about the past or future.

Binomial: Literally means to divide into two numbers. The mode of processing that is based upon the recognition and separation of discrete entities within the whole. Printing is binomial writing. Most computers use binomial processing. Digital is another word commonly used for this.

Block: When Chi is unable to flow. A dam in the normal and healthy vital energy flow.

Centering: A term used in meditation, the martial arts, and spiritual practices meaning to focus attention exclusively in the present tense. It is typically characterized by basic brainwave patterns in the alpha range.

Chi: Also spelled Qi. It is the Chinese concept of the vital life energy. It is comparable to the Prana in the East Indian Vedanta tradition. It is conceived as a dynamic flowing quality. It constantly flows in the body in a closed circuit. Meridians are abstract representations on the surface of the body indicating that general flow pattern. Health is a balanced even flow of Chi. Disruptions or blocks in that flow cause or are disease.

<u>Chakra</u>: In the East Indian tradition energy is perceived to enter the body as a spiraling vortex. They recognize twelve major Chakra feeding directly into the midline or spine and numerous smaller ones distributed through out the rest of the body. This energy is immeasurable by physical means so it corresponds to the implicated energy described in Quantum Mechanics.

<u>Clairaudient</u>: An individual that perceives implicated or subtle energy information as sounds. They often report hearing voices in their head or ear. There is no actual sound or sensory hearing involved.

<u>Clairsentient</u>: An individual that perceives implicated or subtle energy information as "feelings". This may take the form of sympathetic emotions or actual physical sensations such as pain, pressure, or heat.

<u>Clairvoyant</u>: An individual that perceives implicated or subtle energy information visually. They will report actually seeing internal health problems in their minds' eyes. Their eyes and light rays are not actually involved.

<u>Coherent</u> <u>radiation</u>: Energy projected with consistent wavelength and phase. This allows us to capture and exploit energy more efficiently.

<u>Complementary</u> <u>Colors</u>: The color pairs of red-green, yellow-violet, and blue-orange are the complimentary colors. Our eyes naturally balance each of these pairs in much the same way a photographic print is related to its negative. They occur directly across from each other on a spectral color wheel and produce the color brown when mixed.

<u>Contra</u>-<u>lateral</u>: The opposite side of the body. Right is contra-lateral to left.

<u>Cyber</u>-<u>physiology</u>: A term used by Robert Stone to mean controlling physiological mechanisms with the mind. Others refer to this as creative visualization or positive affirmations.

<u>Deductive</u> <u>Reasoning</u>: Reasoning from general rules to specific details. It is the basic form of reasoning used by the analog mode.

<u>Delta</u> <u>Waves</u>: EEG brainwave frequencies less than 4 hertz. This low pitched state of mind is associated with dreamless sleep, deep meditation, and unconsciousness.

<u>Digital</u>: Another word for binomial processing. Digital technologies typically are used to accentuate the differences or sharpen the edges in reality.

<u>Divine</u>: Same as dowse. Defined in chapter of same name.

<u>Enlightenment</u>: Reliance upon reason and experience as opposed to dogma and tradition. Tension is created in life when our logic and reasoning are out of sync with reality. This tension is created by our unsuccessful efforts to impose or force an inappropriate format upon reality. Thinking in terms of <u>should</u> is typical of the unenlightened. Enlightenment reduces tension in life (healing) as the individual learns to go with the flow of reality rather than resisting it. Accepting the laws of a proven logical format like Quantum Mechanics is enlightenment.

<u>Grounding</u>: Often equated with centering. It is the psychological state that results from slowing down the basic brainwave frequency and taking control of the mental processes. Grounded individuals generally display less tension and appear comparatively calm and quiet.

<u>Homeopathy</u>: Literally means *same disease*. A medical treatment protocol that is based upon treating a symptom with the subtle energy of a physical agent that would cause a similar problem in a healthy

normal individual. Skin rashes are treated with the energy of poison ivy, for example.

Homeopathic Remedy: The subtle energy profile of a potential physical poison is separated from its original chemical manifestation and transferred to an innocuous substance like water for administration. This was done historically with a series of dilutions and sucussions. Radionic machines can also be used to make this type of remedy.

Incoherent radiation: Energy projected with different wavelengths and phases. With the present technology, we have a difficult time capturing and exploiting this type of energy with any degree of efficiency.

Inductive Reasoning: Reasoning from specific details to generalizations. It is the basic form of reasoning used by the binomial mode.

Intoning: Producing a monotonous tone or note with the voice with specific intent. The sound vibrations of different vowels or syllables vibrate specific physical structures in the body. These vibrations can be used to disassociate pain frequencies or stimulate physiologic activities such as hormone production.

Ipsi-lateral: The same side of the body.

Koans: Carefully devised nonsensical riddles which are meant to demonstrate, for the student of Zen, the limitations of logic and language.

Linear: In line. One thing following another.

Logic: Reasoning. Laws defining how the different elements or subsets of a dimension of reality are related to each other.

Manifesting: Revealed or becoming apparent to the senses.

Mantra: A phrase or statement repeated continuously during meditation or prayer to help avoid distraction and maintain the alpha state of consciousness. They are typically repeated as rote chants or songs without literal cognition.

Meditation: The act of purposely slowing down brainwave activity. This is generally believed to be the way to achieve enlightenment, heal, and increase intuitive insights.

Meridian: Term used to denote the normal routes chi flows through the body. It is more abstract than physical.

<u>Miasm</u>: In homeopathy it is a tendency for a disease pattern to reoccur. It is generally recognized as some latent effect of a specific disease condition suffered by the previous generation. We know today this is a result of genetic transference. Scientists are using transference today in genetic engineering. To their credit, homeopaths identified this mechanism before science discovered genes.

<u>Mode</u>: A manner or way of acting, doing or being; method or form.

<u>Moxibustion</u>: The practice of stimulating acupuncture points and adjusting the subtle energy of a patient with the heat of combusted materials and herbs. The Chinese primarily use the herb Mugwort for this. It is considered a very effective and important technique in China. Many of the early texts were titled Acupuncture and Moxibustion.

<u>Newtonian Physics</u>: A comprehensive physical science developed by Isaac Newton. It is widely accepted and taught as the most practical and realistic interpretation of physical reality. It is the logical format most commonly used by rational western society and consequently determines its world view. It begins with the premise; "Reality is

made up of separate and immutable matter and energy." The only energy recognized and described in this paradigm is explicated or measurable.

Octave: The spectrum of energy wavelengths between resonant lengths. There are seventy octaves total in the electromagnetic spectrum of wave lengths. Visible light is the 49th octave with tiny wavelengths between 3500 and 7000 angstroms long. There are several octaves between the 4th and 15th with wavelengths meters long that we perceive as sound. A musical octave (eight notes) is the six sequential tones between two pitches of the same note (resonant frequencies).

Paradox: Contradictory logical conclusions.

Perturbation: Adjustments made in a logical format to make inaccurate conclusions more compatible with reality. Perturbation adjustments always compromise and complicate the original basic logical format.

Potency: The potency of a classic homeopathic remedy indicates how many times it was diluted and shaken. High dilution equals high potency. Experience indicates that the higher the potency the more powerful the healing and the more specific the effect.

Pythagoras's Comma: The Pythagorean philosophy attempted to equate mathematical and geometrical forms with the basic laws of nature. This movement took on religious overtones with ideas about the sacredness of numbers like seven, twelve and the "Music of the spheres." The fly in the ointment proved to be the fact that the recognized musical notes could not be fit exactly into an octave frequency circle. This "comma" or defect was eventually translated as proof of man's biblical fall from perfection.

Qi Gong: The Chinese protocols that actively use visualizations and intent to adjust the flow of the life force in the body. It literally translates as "Working with or moving Qi." It makes use of physical exercises and hand movements.

Quanta: Energy units. In subatomic reality, implicated energy transforms into explicate energy for us in discrete uniform steps or stages. These stages correspond to, and therefore are defined by, the different orbits of subatomic particles in the physical model of reality. Light energy is transferred as a "photon", for example, and all photons are equal <u>quantitatively</u>. A photon is the unit of energy necessary to push an electron into the next higher orbit. If we measure the amount of water in the river with a cup, it appears obvious to us that water occurs in nature in cup units.

<u>Quantum</u> <u>Field</u>: A name generally explaining the mechanism of entanglement or connectedness described in Quantum Mechanics. Ether, Higgs field, universal consciousness, or Morphic fields are all names used at one time or other by different individuals to describe what exactly is waved or vibrated in the transmission of electromagnetic radiation or vibrated in implicated energy. It is the medium which transfers effects. In this book the quantum field and implicated energy are equated.

<u>Quantum</u> <u>Mechanics</u>: The science of reality that defines reality as totally energetic. It was developed by subatomic scientists and has proved to be the most accurate and productive logical format ever developed by man. Reality is described in terms of interfering wave functions.

<u>Radionics</u>: A divining/treatment modality developed by the British. It began as a form of pendulum divination and evolved over the years

into more formalized radionic machines that are based upon divining energy frequencies with a finger stick plate. The machine allows the practitioner to keep track of and hold on to divined information on a set of dials. The machines are also able to produce homeopathic-like remedies from stock samples. They are used to broadcast treatment frequencies like a radio transmitter over distance to a patient. I believe, however, that the machines simply act as conduits for the practitioner's conscious intent, which we know has this ability.

<u>Reaction</u>: An automatic conditioned reflex. It is triggered by circumstances and is invariable in character and degree. It is an example of binomial mode processing.

<u>Respond</u>: Rational graded activities designed to achieve a particular goal or purpose in a given situation.

<u>Reiki</u>: A Japanese healing modality that was developed by Dr. Mikao Usui. It is based upon manipulating the energy of the body with a combination of physical and mental techniques. Energy is adjusted with static physical touch, visualizations, and long distance projections. It has many aspects that are similar to religious hands on

healing practices. It was originally brought to this country as a type of pyramid business scheme.

Resonance: Intensification of effect. Sympathetic vibration. Wavelengths that are one or two multiples of each other reinforce each other. Beyond 3x the effect is minimal.

Science: The logical exploration of reality. Different logics produce different sciences.

Shu Points: These are acupuncture points that have traditionally been used to diagnose the condition of a whole meridian or part of the physical body. In Veterinary Acupuncture a series of Bladder meridian points along the spine have been used traditionally to evaluate the over-all balance of the major meridians. A painful response to palpation or an obvious drop in relative electrical resistance indicates problems.

Silva Method: Jose Silva developed a method of treatment that relies upon some self hypnosis techniques. His system primarily involves some practical and useful positive visualization techniques. The great success of this system is based upon his realization that healing occurs only when brain wave activity assumes one of the slower EEG patterns. He identifies that condition as the Alpha brain wave level.

Stick Plate: A diving devise based upon the electro-conduction of the skin. The tips of the fingers are rubbed in a circular pattern over the slick surface of a ceramic tile. The finger continues to slide when the analog answer is "no" and stick suddenly when the answer is "yes." Quantitative or numbered information can be dowsed by combining the stick plate with a calibrated analog dial.

Subtle Energy: Refers to the physically immeasurable energy of the higher dimensions of reality. Used by many to refer to qualities of the life process generally unrecognized by conventional physical sciences. Quantum physics calls it implicated energy.

Super-string Theory: A scientific logical format that describes reality in terms of fundamental vibrations rather than waves. The vibrating strings are closed circles or visualized as vibrating spheres of influence. The theory is still very much a work in progress.

Surrogate treatments: The energetic treatment of a patient by the manipulation of another individual who is focused consciously upon the patient.

<u>Tao</u>: Taoist word for the way things work.

<u>Theta</u> <u>Waves</u>: EEG brainwave frequencies in the 4 to 8 range. This state of mind is associated with deeper levels of prayer or meditation. Semiconscious is a word often used to describe it. Some say that it is associated with super learning abilities and dreaming sleep.

<u>Therapeutic</u> <u>Touch</u>: A system of medical treatment that attempts to manipulate the energy of the body with hand movements. Perceived energy blocks or improper distributions are waved, pushed or drug into place with distant passes of the hands through the Aura. There is actually no physical touching in this protocol. It is widely taught to nurses in this country. It was originally developed by DoLores Krieger.

<u>Tai</u> <u>Chi</u>: The Chinese practice of moving meditation. It can be difficult to clear the mind of the busy Beta wave thought patterns. Concentrating on a precise, slow, introspective movement protocol occupies the brain's focus, helping to first obtain and then maintain the slower Alpha wave activity needed for effective meditation and prayer.

<u>Ting</u> <u>Points</u>:   The terminal point of each of the twelve major meridians on the digits and toes. Half are the first numbered point of a meridian and half are the last numbered point. They are commonly used in equine acupuncture as diagnostic points. Relative puffiness is an indication of problems.

<u>Voll</u>:   A diagnostic and treatment system developed by the German Acupuncturist Reinhold Voll.   It is based upon evaluating the measured electrical resistance of the ting points. The balancing effect of potential bottled remedies is then tested in a trial and error format until an effective one is found, and it is taken as a homeopathic. Voll is one of the most logical systems in general use today and has recently been successfully interfaced with computers in several different applications.

<u>Waveform</u>:   Energy that manifests as a replicating pattern of amplitude fluctuations. The effect it has in physical reality depends upon the consistency of that specific pattern of fluctuations and the time/distance it takes to manifest (wavelength). Uniform or coherent waveforms produce powerful effects. The strength of the effect is proportional to the amplitude of the waveform. The right frequency

projected with sufficient amplitude can shatter a crystal wineglass or kidney stone.

<u>Wavelength</u>: The distance it takes to complete one replication of a uniform waveform projection. The wavelength determines what elements in physical reality will absorb and ultimately be affected by it. The length of an antenna determines what frequencies it will absorb and transmit to its receiver, for example. The wavelengths absorbed by the skin you perceive as heat.

<u>Yoga</u>: A Hindu meditation protocol utilizing controlled breathing and precise static postures to maintain the necessary Alpha brainwave pattern.

<u>Zero Balancing</u>: A passive method of energy balancing based upon the idea that the body intuitively knows and seeks a healthy balance. It also recognizes that actively doing something compromises the Alpha state which is required for healing or balancing. The fulcrum is a powerful tool that does nothing. In zero balancing the therapist simply provides a static point or fulcrum for the patient to balance over.

ABOUT THE AUTHOR

I am a 1969 graduate of the College of Veterinary Medicine, Iowa State University, in Ames, Iowa. There I worked as a professional medical illustrator and instructor in the anatomy department under Doctor Robert Getty. I moved west to facilitate my pastime of horse and mule packing.

For the next thirty years, I practiced conventional veterinary medicine in Utah and Idaho. The emphasis of my practice in the last twenty of those years has been contract dairy herd health work. Herd health programs are more comprehensive and holistic by nature than ambulatory work. Concerns about drug residues ultimately led me to investigate and use other alternative medical approaches. In 1992, I became a certified veterinary acupuncturist under the tutelage of the father of American Veterinary Acupuncture, Doctor Grady Young. Doctor Young also introduced me to the modality called Radionics. Neither proved to be the silver bullet I was hoping for. An investigation of several other alternative modalities followed. I found each of the modalities I studied to be effective, but they were impossible to integrate.

Subsequent studies of theoretical physics, logic, and neurophysiology provided a true epiphany and solution. I discovered that all the health care modalities I had studied could be logically integrated into one comprehensive, effective, and simple holistic format.

This simple and basic logical approach has proved to be spectacularly successful in dealing with difficult medical conditions like allergies and cancer. It also led to the discovery of the unique treatment protocol I call "Sound Therapy".

Printed in the United States
1066500005B/118-138